CRC Handbook of Nucleobase Complexes

Volume I
Transition Metal Complexes of Naturally Occurring Nucleobases and Their Derivatives

Editor

James R. Lusty
Head of Department
School of Chemistry
Lancashire Polytechnic
Preston, England

CRC Press
Boca Raton Ann Arbor Boston

Library of Congress Cataloging-in-Publication Data

CRC handbook of nucleobase complexes/editor, James R. Lusty.
p. cm.
Includes bibliographical references.
Contents: Vol. 1. Transition metal complexes of naturally occurring nucleobases and their derivatives.
ISBN 0-8493-3281-8 (v. 1)
1. Nucleotides. 2. Nucleosides. 3. Purines. 4. Pyrimidines. 5. Transition metal complexes. I. Lusty, James R., 1951—
[DNLM: 1. Base Sequence. 2. Nucleosides. 3. Nucleotides. QU 58 C911]
QP625.N89C73 1990
547.7'9—dc20
DNLM/DLC
for Library of Congress 89-70892
CIP

Direct all inquires to CRC Press, Inc., 2000 Corporate Blvd., N.W., Boca Raton, Florida, 33431.

International Standard Book Number 0-8493-3281-8 (v. 1)

Library of Congress Card Number 89-70892
Printed in the United States

PREFACE

This preface is necessarily brief since this work consists of a series of self-contained sections, each giving general comments on the scope of work covered and a short introduction. Some of these are comprehensive while other authors have preferred to refer the reader to specific review articles.

It has been the aim of this series to document fully the range of transition metal complexes that have been prepared using the purine and pyrimidine bases and their derivatives.

The nucleobases are divided into two groups based on either the six-membered pyrimidine ring or the nine-membered fused rings of the purine bases. Within each group the sections are further divided into the nucleobases, the nucleosides, and the nucleotides and the oligonucleotides. Within each section the major naturally occurring bases and some of their derivatives are listed. Each derivative is indexed prior to each table.

All transition metals are covered, including, where appropriate, the lanthanides and actinides. The metals associated with the transition block, zinc, cadmium, and mercury, are also documented as they contribute an important part of the spectroscopic and structural data that is currently available.

Such is the rate at which chemists and biochemists work and record their findings, that many new complexes will have been synthesized before this book is published.

The work is divided into two volumes, the first of which lists the complexes and the methods used to study them. The second volume gives details of the spectroscopic properties of a wide range of complexes using a variety of techniques. It is meant to demonstrate the range of methods that can be used and their application to structure elucidation.

The Editor would wish to thank the numerous contributors who have been so obliging and helpful during every stage of the development of this book. Additional thanks go to the editorial staff at CRC Press, especially James McCabe, Amy Skallerup, Sandy Pearlman, and James Brody. Finally, special thanks and gratitude must be recorded to Carole Knight and Jean Moffatt who have unselfishly slaved away with the typing of the entire work.

Notification of any omissions, any errors in the data, any corrections or criticisms will be welcomed in order to improve the value and scope of this work.

CONTRIBUTORS

Juan R. Alabart, Ph.D
Research Associate
Department of Chemistry
University of Barcelona
Tarragona, Spain

Hardy Sze On Chan, Ph.D.
Senior Lecturer
Department of Chemistry
National University of Singapore
Republic of Singapore

Kenji Inagaki, Ph.D.
Lecturer
Faculty of Pharmaceutical Sciences
Nagoya City University
Nagoya, Japan

Badar Taqui Khan, Ph.D., F.N.A.Sc.
Professor Emeritus
Department of Chemistry
Osmania University
Hyderabad, India

Yoshinori Kidani, Ph.D.
Professor
Faculty of Pharmaceutical Sciences
Nagoya City University
Nagoya, Japan

Bernhard Lippert, Dr.rer.nat.habil.
Professor
Department of Chemistry
University of Dortmund
Dortmund, West Germany

James R. Lusty, Ph.D.
Head of Department
School of Chemistry
Lancashire Polytechnic
Preston, England

Virtudes Moreno, Ph.D.
Professor
Department of Chemistry
University of Barcelona
Tarragona, Spain

Angel Terrón, Ph.D.
Associate Professor
Department of Chemistry
University of the Balearic Islands
Palma de Mallorca, Spain

Peter Wearden, Ph.D.
Senior Lecturer
Department of Chemistry
Lancashire Polytechnic
Preston, England

To Jackie,
Matthew, Nicola, and Helen

TABLE OF CONTENTS

NOTES FOR GUIDANCE

It has been a difficult decision to decide on a standard format of recording complexes. In many cases original authors have used nonstandard abbreviations, but hopefully most of these have now been removed in this text. While it has not been recorded by all authors, it has been assumed that elemental analysis or some form of microanalysis on the complexes has been performed. In most cases, C, H, and N together with S or halogen analysis have been undertaken, while some papers have also recorded metal analysis by spectroscopic, titrimetric, or gravimetric methods.

Solution studies have been included where contributors felt these studies have added to the scientific knowledge base and their authors have demonstrated that a rigorous approach to species identification has been followed.

In the formula listings, water of crystallization has been mainly excluded, unless it is essential for the stereochemistry of the complex or is part of a definitive study such as X-ray diffraction.

Some polymeric species are ill-defined, and the policy has been to record the single unit where possible. A large number of complexes, particularly those of the purine-1-oxides, are recorded as possible polymeric species, and the reader is cautioned as to the exact nature of the complex reviewed.

Another problem arises where a preliminary communication records a different entry to the main work which subsequently follows. In most cases both studies are recorded.

A list of abbreviations and methods of study follow.

METHODS OF STUDY

atta	antitumoral activity
calc	M.O., SCF calculations
chr	chromatography
CD	circular dichroism
cv	cyclic voltametry
cond	conductimetric measurements
ech	electrochemical measurements
eph	electrophoresis
esr	electron spin resonance spectroscopy
EXAFS	extended X-ray photoelectron spectroscopy
hch	hypochromic spectra
HPLC	high-performance liquid chromatography
ion	ion exchange
ir	infrared spectroscopy
kin	kinetic studies
ks	stability constant data, data relating to ΔHf
mag	magnetic measurements

MB	Mossbauer
MD	Magnetic dichroism
mol	molecular weight determination
ms	mass spectrometry
nmr	nuclear magnetic resonance spectroscopy
OR	optical rotation measurements
pK	pK data
pol	polarography
pot	potentiometric titrations
ra	radioactivity studies
ram	Raman spectroscopy
sfk	stopped-flow technique
spec	spectroscopy techniques (various)
therm	thermal studies
thermodyn	thermodynamic measurements
titr	titrimetric measurements
XPS	X-ray photoelectron spectroscopy (esca)
X-ray	X-ray techniques
uv	ultraviolet spectroscopy and/or UV/VIS spectroscopy

KEY OF ABBREVIATIONS

ac	acetate
acac	acetylacetonate
ade	adenine
adelox	adenine-*N*(1)-oxide
6A13dme3fura	6-amino-1,3-dimethyl-3-formyluracil
6A13dme5fura	6-amino-1,3-dimethyl-5-formyluracil
6A13dme5NOura	6-amino-1,3-dimethyl-5-nitrouracil
ado	adenosine
adolox	adenosine-*N*(1)-oxide
ε-ado	1,*N*(6)-ethenoadenosine
ADP	adenosine-5′-diphosphate
5A6HOcyt	5-amino-6-hydroxycytosine
5A6HO2Scyt	5-amino-6-hydroxy-2-thiocytosine
6A2Scyt	6-amino-2-thiocytosine
6Aeapur	6-aminoethylaminepurine
6A5fura	6-amino-5-formyluracil
6A5f1meura	6-amino-5-formyl-1-methyluracil
6A5f3meura	6-amino-5-formyl-3-methyluracil
2Ameguo	2-aminomethylguanosine
6A1me5NOura	6-amino-1-methyl-5-nitrouracil
6A3me5NOura	6-amino-3-methyl-5-nitrouracil

2A9mepur	2-amino-9-methylpurine
8A9Mepur	8-amino-9-methylpurine
6A5NOura	6-amino-5-nitrouracil
2A6Spurr	2-amino-6-thiopurineriboside
6A2Sura	6-amino-2-thiouracil
ala	alanine
AMP	adenosine-5′–monophosphate
asp	asparagine
ATP	adenosine-5′-triphosphate
8azade	8-azaadenine
6azura	6-azauracil
bipy	2,2′-bipyridine
bdppe	1,2-bis(diphenylphosphine)ethane
bn	2,3-diaminobutane
bpe	1,2-bis(pyridin-2-yl)ethane
8Brguo	8-bromoguanosine
5Brura	5-bromouracil
Bu	Butyl
Bu_3P	tri-*n*-butylphosphine
bz	benzyl
5bzcys	5-benyl-L-cysteine
9bz6Spur	9-benzyl-6-thiopurine
caf	caffeine
chbma	1,1-bis(methanamine)cyclohexane
2Cl9mepur	2-chloro-9-methylpurine
8Cl9mepur	8-chloro-9-methylpurine
5Cl1meura	5-chloro-1-methyluracil
5Clura	5-chlorouracil
CMP	cytidine-5′-monophosphate
3CMP	cytidine-3′-monophosphate
COD	1,5-cyclooctadiene
CTP	cytidine-5′-triphosphate
cyd	cytidine
cys	cysteine
cyt	cytosine
dab	*o*-phenylenediamine
dach	1,2-diaminocyclohexane
1,3dach	1*R*,3*S*-diaminocyclohexane
dad	diaminodiol; 2,3-diamino-2,3-dideoxy-D-threitol
dado	2′-deoxyadenosine
26dApur	2,6-diaminopurine
56dA2Scyt	5,6-diamino-2-thiocytosine
dat	3,4-diaminotoluene
date	diaminotetrol; 2,4-diamino-3,4-dideoxy-D-iditol

datr	diaminotriol; 2,3-diamino-2,3-dideoxy-D-xylitol
56dAura	5,6-diaminouracil
dazado	deazaadenosine (Tubercidin)
dazino	deazainosine
dcyd	2′-deoxycytidine
13det5Fura	1,3-diethyl-5-fluorouracil
dguo	2′-deoxyguanosine
56dHdHO1methy	5,6-dihydro-5,6-dihydroxy-1-methylthymine
56dHdHOthy	5,6-dihydro-5,6-dihydroxythymine
56dH1meura	5,6-dihydro-1-methyluracil
56dHura	5,6-dihydrouracil
dien	diethylenetriamine
DIPSO	di-isopropylsulphoxide
dino	2′-deoxyinosine
dma	dimethylacetamide
dmdap	2,2-dimethyl-1,3-diaminopropane
29dmeade	2,9-dimethyladenine
79dmeade	7,9-dimethyladenine
89-dmeade	8,9-dimethyladenine
dmeado	*N*(6),*N*(6)-dimethyladenosine
13dme5Fura	1,3,-dimethyl-5-fluorouracil
dmeguo	*N*(2),*N*(2)-dimethylguanosine
dmehyp	7,9-dimethylhypoxanthine
dmen	*N*,*N*′-dimethyl-1,2-diaminoethane
dmeprgua	*N*(2),*N*(2)-dimethyl-9-propylguanine
29dmepur	2,9-dimethylpurine
69dmepur	6,9-dimethylpurine
89dmepur	8,9-dimethylpurine
19dme6Spur	1,9-dimethyl-6-thiopurine
29dme2Spur	2,9-dimethyl-2-thiopurine
13dmeura	1,3-dimethyluracil
13dmexan	1,3-dimethylxanthine
38dmexan	3,8-dimethylxanthine
dmf	dimethylformamide
dmg	dimethylglyoxime
dmopda	4,5-dimethyl-*o*-phenylenediamine
DMSO	dimethylsulfoxide and related ligands
dmtn	*N*,*N*′-dimethyl-1,3-diaminopropane
doro	diorotic acid
dpen	1,2-diphenylethylenediamine
DPPH	diphenylpicrylhydrazyl radical
13dpr5Fura	1,3-dipropyl-5-fluorouracil
dSura	2,4-dithiouracil
dthd	2′-deoxythymidine

dthy	dithymine
durd	2′-deoxyuridine
3ecma	3-((ethoxycarbonyl)methyl)-adenine
en	ethylenediamine; and 1,2-diaminoethane
EOA	ethanolamine
et	ethyl
5etcys	5-ethyl-L-cysteine
9etgua	9-ethylguanine
8et3mexan	8-methyl-3-methylxanthine
8etthp	8-ethyltheophylline
8etxan	8-ethylxanthine
5Fura	5-fluorouracil
GLP	β-glycerophosphate
gly	glycine
glyala	glycylalanine
glyasp	glycylasparagine
glygly	glycylglycine
glyhis	glycylhistidine
glyphe	glycylphenylalanine
glytyr	glycyltyrosine
3GMP	guanosine-3′-monophosphate
GMP	guanosine-5′-monophosphate
GMPme	guanosine-5′-phosphatemonomethylester
GTP	guanosine-5′-triphosphate
gua	guanine
guo	guanosine
his	histidine
5HCl6OHCl1meura	5-hydrochloro-6-hydroxychloro-1-methyluracil
his	histidine
hisam	histamine
6HOeapur	6-hydroxyethylaminepurine
6HOemapur	6-hydroxyethylmethylaminepurine
6HO2Scyt	6-hydroxy-2-thiocytosine
hyp	hypoxanthine
hypbu	1,4-bis(hypoxanth-9-yl)butane
hyppo	1,3-bis(hypoxanth-9-yl)propanol
hyppr	1,3-bis(hypoxanth-9-yl)propane
ile	isoleucine
IMP	Inosine-5′-monophosphate
imz	imidazole
ino	inosine
ipa	isopropylamine
8iprthp	8-isopropyltheophylline
ipentado	*N*(6)-(Δ^2-isopentyl)adenosine

ipr	isopropane
ipro	isopropanol
isopado	2′,3′-*O*-isopropylideneadenosine
isopguo	2′,3′-*O*-isopropylideneguanosine
isopino	2′,3′-*O*-isopropylideneinosine
isopRMP	2′,3′-*O*-isopropylidene-β-D-ribofuranosyl-6-mercaptopurine
5Iura	5-ioduracil
leu	leucine
me	methyl
3meade	3-methyladenine
9meade	9-methyladenine
1meado	1-methyladenosine
6meado	*N*(6)-methyladenosine
5mecys	5-methyl-L-cysteine
1mecyt	1-methylcytosine
1me2dguo	1-methyl-2′-deoxyguanosine
9megua	9-methylguanine
1meguo	1-methylguanosine
7meguo	7-methylguanosine
1meino	1-methylinosine
7meino	7-methylinosine
2meo9mepur	2-methoxy-9-methylpurine
8meo9mepur	8-methoxy-9-methylpurine
6meopr	6-methoxypurineriboside
6meoro	6-methylorotic acid
9mepur	9-methylpurine
9me6Spur	9-methyl-6-thiopurine
6me2Sura	6-methyl-2-thiouracil
met	methionine
1me5NOura	1-methyl-5-nitrouracil
1methy	1-methylthymine
1meura	1-methyluracil
6meura	6-methyluracil
1mexan	1-methylxanthine
3mexan	3-methylxanthine
7mexan	7-methylxanthine
8mexan	8-methylxanthine
9mexan	9-methylxanthine
7mexao	7-methylxanthosine
mit	1-methylimidazol-2-thiol
NMeN′SEN	*N*-methyl-*N*′-salicylideneethylenediamine and related ligands
5NObzcys	5-nitrobenzyl-L-cysteine

5NOoro	5-nitroorotic acid
5NOura	5-nitrouracil
nta	nitrilotriacetic acid
oct	octyl
Oetcys	*O*-ethyl-L-cysteine
Omeocyd	*O*-methoxycytidine
Omecys	*O*-methyl-L-cysteine
opda	*o*-phenilenediamine
oro	orotic acid
ox	oxalate
pa	propylamine, aminopropane
pen	1-phenylethylenediamine
8pethp	8-pentyltheophylline
Ph	phenyl
Phe	phenylalanine
phen	1,10-orthophenanthroline
PHMB	parahydroxymercurybenzoate
8phthp	8-phenyltheophylline
pic	2-methylpyridine
pmdien	1,1,4,7,7-pentamethyldiethylenetriamine
pmt	pentamethylenetetrazole
pn	1,2-diaminopropane
pro	proline
8prthp	8-propyltheophylline
pur	purine
purlox	purine-*N*(1)-oxide
py	pyridine
pym	pyrimidine
rfpur	[9-(β-D-ribofuranosyl)purine]
5Rura	5-*R*-uracil(alkyl), 5-alkyluracil
6Segua	6-selenoguanine
ser	serine
6Sgua	6-thioguanine
8Sgua	8-thioguanine
6Sguo	6-thioguanosine
8Sguo	8-thioguanosine
Spur	thiopurine
6Spur	6-thiopurine
6Spurr	6-thiopurineriboside
SSA	sulfosalicyclic acid
2Sura	2-thiouracil
2SSura	*S*,*S*-2-thiouracil
4Sura	4-thiouracil
2Sxan	2-thioxanthine

t6A	*N*-[9-(β-D-ribofuranosyl)purin-6-ylcarbamoyl)-threonine
tba	tributylamine
tct	tricanthine, 3-(γ,γ-dimethylallyl)adenine
tea	triethylamine
teaado	tetracetyladenosine
thb	theobromine
thd	thymidine
thp	theophylline
thf	tetrahydrofuran
thr	threonine
tmdap	2,2,*N*,*N*,-tetramethyl-1,3-diaminopropane
tmegua	*N*(2),*N*(2)-dimethyl-9-methylguanine
tmen	*N*,*N*,*N*′,*N*′-tetramethyl-1,2,diaminoethane
8tmexan	1,3,8-trimethylxanthine
9tmexan	1,3,9-trimethylxanthine
TMP	thymidine-5′-monophosphate
TMS	tetramethylsilane
tmtn	*N*,*N*,*N*′,*N*′-tetramethyl-1,3-diaminopropane
tn	trimethylenediamine
traado	2′,3′,5′-triacetyladenosine
traguo	2′,3′,5′-triacetyliguanosine
traino	2′,3′,5′-triacetylinosine
trp	tryptophan
ttha	triethylenetetraminehexaacetate
TTP	thymidine-5′-triphosphate
tu	thiourea
UMP	uridine-5′-monophosphate
UTP	uridine-5′-triphosphate
ura	uracil
urd	uridine
val	valine
xan	xanthine
xao	xanthosine
XMP	xanthosine-5′-monophosphate

Section 1

TRANSITION METAL COMPLEXES OF PYRIMIDINE NUCLEOBASES AND THEIR DERIVATIVES

Bernhard Lippert

INTRODUCTION

Interest in metal binding properties of nucleic acids constituents or models of varying complexity has substantially increased over the last 2 decades. Findings that metals can have a pronounced effect on nucleic acids structures and, hence, can be expected to affect chromation function has greatly stimulated activities in this area.[1] Rosenberg's discovery of the antineoplastic activity of certain platinum coordination complexes in the late 1960s and the present clinical success of Cisplatin have provided additional impetus in this direction.[2,3] Other relevant aspects include, for example, possible applications of metal complexes as chemical probes for nucleic acid structures, specific stains for nucleic acids, or tools for separating nucleic acids of different compositions.[4-6]

The understanding of basic principles of metal-nucleic acids interactions is, for obvious reasons, of great significance for a proper understanding of biological effects.[7] Systematic studies of metal complexes of simple models of nucleic acids, frequently supported by X-ray structural work,[8] are substantially contributing to this goal. Among metal-nucleobase complexes, compounds containing the transition element platinum in its +2 oxidation state have been particularly prevalent.[9-11]

As far as purine vs. pyrimidine bases and their respective significance as donors for transition metals are concerned, it appears that, at least with DNA, the N7 site of purines (guanine, adenine), due to basicity, availability at neutral pH, and good accessibility in duplex DNA, is more important than any of the other sites.

PYRIMIDINE NUCLEOBASES

The common pyrimidine nucleobases present in natural nucleic acids are uracil (RNA), thymine (DNA), and cytosine (RNA, DNA). There are many additional minor pyrimidine nucleobases present temporarily in DNA (e.g., 5-methyl-cytosine) or permanently in tRNAs.[12] Derivatives of the naturally occurring pyrimidine nucleobases represent an important class of antitumor and antiviral agents, examples being 5-fluorouracil or 3′-azidothymidine, respectively. The pyrimidine ring is, apart from nucleobases, also found in other physiologically relevant molecules such as vitamins (e.g., vitamin B_1), antibiotics (e.g., bleomycin), and narcotics (e.g., barbiturates). In some instances, an

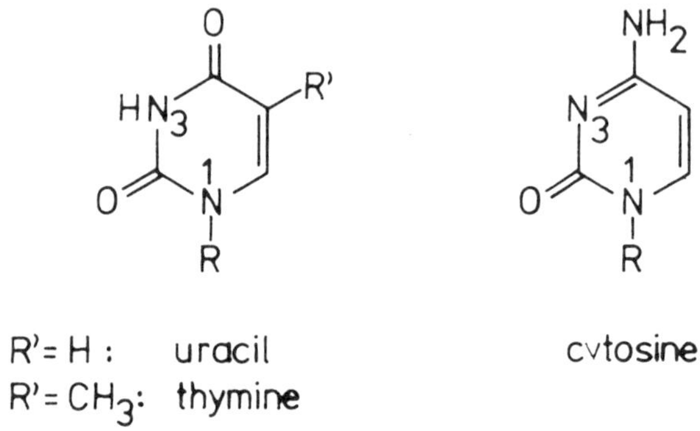

FIGURE 1. Atom numbering scheme of uracil (thymine) and cytosine. R + H: unsubstituted bases; R + CH_3: 1-methyl derivatives; R = ribose (deoxyribose):nucleosides; R = ribose (deoxyribose):nucleotides.

involvement of transition metal ions in the biochemistry of these compounds has been suspected or even explicitly proven, as with the bleomycins.

Figure 1 depicts the predominant tautomeric structures and the numbering scheme of uracil (thymine) and cytosine. In the nucleic acids, the N1 position is bound to a ribose (RNA) or a deoxyribose (DNA) moiety, which in turn is connected with a phosphate residue which links the 3′ and 5′ positions of adjacent bases. As far as metal binding properties are concerned, the free nucleobases (R = H) do not represent ideal models of the respective nucleosides (R = ribose or deoxyribose) and nucleotides (R = ribose phosphate or deoxyribose phosphate) because of possible metal coordination at a site (N1) unavailable in nucleic acids. Still, the parent bases have been frequently applied in metal binding studies.

Depending on the pH and the medium, the free nucleobases can exist in different protonation states. In aqueous solution, uracil and thymine exist as neutral species, as monoanions or dianions, depending on pH. In super acidic medium, also mono- (pK_a ≃3) and dications are formed.[13] Cytosine exists as a cation (N3 protonated), a neutral molecule, and an anion (N1 deprotonated) in the normal pH range. In strongly acidic medium, further protonation to give the dication, is expected. In strongly basic medium, deprotonation of the exocyclic amino group can occur. As far as physiological pH conditions are concerned, only neutral forms of uracil, thymine, and cytosine are *a priori* expected to play a role as metal-coordinating ligands. However, the pH criterion has limitations in that metal binding to sites unexpected from acidity consideration has been observed in serveral instances (*vide infra*). In these cases, the metal electrophile has substituted very weakly acidic protons. It appears, that the existence of metal-hydroxo species at physiological or even acidic pH in many cases enables nucleobase deprotonation and simultaneous metal binding at pH values well below those expected from pK_a data. Moreover, primary metal binding to a neutral nucleobase may sufficiently acidify another proton and facilitate metal

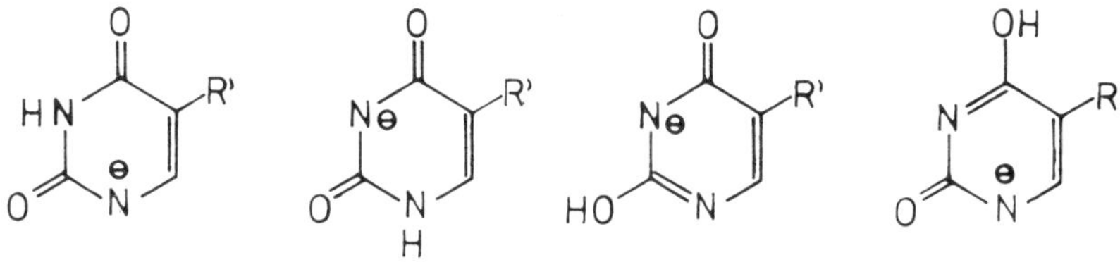

FIGURE 2. Possible tautomer forms of uracil (thymine) monoanion. Only single mesomeric structure is given for each tautomer.

coordination at such a site (e.g., N3,N4 metal binding in 1-substituted cytosines or N3,N1 metal binding in unsubstituted cytosines).

Finally, the concept of H acidity as a measure for metal binding to nucleobases requires modification for base-paired oligonucleotides. Recent findings on the existence of protonated cytosine nucleobases in non-Watson-Crick base pairing schemes,[14] like formation of a triple-stranded H-DNA[15] containing protonated bases, and the long-known existence of poly(C)·poly(CH)$^+$ up to pH 7, support this view.

URACIL AND THYMINE: NEUTRAL FORM, MONOANION, AND DIANION

Unsubstituted uracil (thymine) in its neutral form can exist in six tautomer forms. One of them, the pyrimidine-2-4(1*H*,3*H*)-dione form, strongly exceeds any of the other five forms, as evident from structural, spectroscopic, and theoretical studies.[16-18] Ring substitution and environmental effects (solvent, temperature) may influence the tautomer distribution, however.

Deprotonation of uracil (thymine) according to

$$LH_2 \underset{+H^+}{\overset{-H^+}{\rightleftharpoons}} LH^- \underset{+H^+}{\overset{-H^+}{\rightleftharpoons}} L^{2-}$$
$$pK_{a1} \qquad\qquad pK_{a2}$$

takes place with $pK_{a1} \simeq 9.6$ (uracil), 9.9 (thymine), and pK_{a2} 14.2 (uracil) and 13 (thymine).[19] The monoanion LH^- can exist in four tautomeric structures (Figure 2). From UV spectroscopic studies and comparison with the monoanions of 1- and 3-methyluracil it has been concluded that I and II are the predominant forms present in H_2O solution approximately in a 1:1 ratio,[20, 21] as confirmed by a Raman study.[22] A crystal structure analysis of tautomer I (potassium salt)[23] and a theoretical study on the charge distribution in forms I and II[24] have been reported.

Metal binding can, in principle, take place with any of the four tautomeric forms of LH^-. Crystallographic evidence for the existence of both N1[25] and N3[26] metalated LH^- species has been provided by Mutikainen et al., together with a detailed analysis of structural parameters.[27] Spectroscopic studies of Pt(II) complexes of uracil[28, 29] and thymine[30] have shown the coexistence of the platinated tautomers I and II.

With the uracil (thymine) dianion L^{2-}, tautomerism no longer is possible, but possibilities of metal coordination are numerous. Clear evidence has been put forward for the following metal binding patterns: monodentate binding to N1 or N3[29,30], bidentate binding through N1 and N3,[29,30] tetradentate bridging through N1,O2 and N3,O4[31], and tetradentate chelation (N1 and O2, N3 and O4).[32]

N1-SUBSTITUTED URACIL (THYMINE): NEUTRAL FORM AND MONOANION

With N1-substituted uracil (thymine), the number of possible tautomers is reduced to three. Again, the 2,4-dioxo form is preferred by far. Metal coordination is expected to occur primarily at the exocyclic oxygens, but stabilization of the rare 4-hydroxo, 2-oxo form through metal binding via N3 is possible as well (*vide infra*).

The anion of N1-substituted uracil (thymine) exists in a single form, with the N3 position deprotonated and the negative charge delocalized essentially over the O4–N3–C2–O2 bonds.[33] Metal binding can take place at either N3, O4, O2, or combinations thereof.

CYTOSINE: NEUTRAL FORM AND ANIONS

Like uracil and thymine, cytosine can exist in six tautomeric forms. As with the former, one tautomer, the 4-aminopyrimidin-2(1*H*)-one, represents the preferred species,[18a] as deduced from spectroscopic[34] and structural[16] data, as well as theoretical calculations.[35] The N3 site is capable of accepting a proton ($pK_a \simeq 4.6$)[19a] and is therefore the most basic site of neutral cytosine. Metal coordination is expected to primarily involve this site. In alkaline solution, the N1 position of cytosine becomes deprotonated ($pK_a \simeq 12.2$),[19a] in strongly alkaline solution the exocyclic amino group is expected to lose a proton as well (pK_a 16.7).[36] Thus, the primary metal binding sites to be considered are N3, N1, and, at high pH, the deprotonated N4 group. In addition, despite low basicity, O2 is yet another potential donor site. The exocyclic amino group has little or no donor properties because of delocalization of the electron lone pair of the amino group into the -electron system of the heterocyclic ring.

N1-SUBSTITUTED CYTOSINE: NEUTRAL FORM AND ANIONS

Blocking of the N1 position by an alkyl group or a sugar (sugar phosphate) moiety reduces the number of possible tautomers to three. As with unsubstituted cytosine, the 4-amino, 2-oxo form predominates strongly.[16,35] pK_a values (protonation of N3, deprotonation of N4) are close to those of the unsubstituted cytosine. As to potential metal binding sites, similar considerations as for unsubstituted cytosine apply (except N1).

METAL-STABILIZED RARE TAUTOMERS AND RARE ANIONS

The distribution of tautomers, like the pH-dependent distribution of species in an acid/base equilibrium, in many cases permits a reasonable guess as to what the metal-coordinated ligand is like. For example, neutral cytosine with its predominant amino, oxo tautomer structure is expected to bind a metal in aqueous solution preferentially through N3, or possibly O2, and 1-methyluracil is expected to bind, in moderately alkaline solution, a metal through N3, or possibly one of the two exocyclic oxygens.

There is a growing list of metal nucleobase complexes, in which the nucleobase is present either (1) in a rare tautomeric form, (2) as an anion not to be expected in the pH range of complex formation, or (3) in a rare anionic form. Examples are given in Figure 3. Metal-stabilized rare tautomers (1) have been realized in the case of the 4-hydroxo, 2-oxo tautomers of 1-methylthymine[37] and 1-methyluracil,[38] while the 4-imino, 2-oxo tautomer of 1-methylcytosine has been observed in a Pt(IV) complex, *trans, trans, trans*-$[Pt(NH_3)_2(mecyto)_2(OH)_2](NO_3)_2$.[39] Both the 1-methyluracil and the 1-methylcytosine complexes have been structurally characterized by X-ray analysis. As to (2), displacement of 4-amino protons by one or even two metal electrophiles, as established for Pt(II),[40] Ru(III),[41] and $CH_3Hg(II)$,[42] would never have been predicted on the basis of pH-dependent species distribution, since in no instance were strongly basic conditions applied (although nonaqueous solvents were used in the case of reactions with methylmercury). Substitution of extremely weakly acidic aromatic protons (3) at the 5 positions of cytosine by Hg(II)[43] and at the 5 positions of uracil by Pt(III)[44] leads, formally, to the generation of rare monoanions (cytosine, uracil) and, with uracil derivatives, also to a monoanion of 1,3-dimethyluracil, to a dianion of 1-methyluracil, and possibly even to a trianion of unsubstituted uracil.

MIXED-NUCLEOBASE, MIXED-METAL, AND MIXED-(METAL, NUCLEOBASE) COMPLEXES

In the large majority of studies on metal coordination to nucleobases, the system is comprised of (1) a single metal and a single nucleobase; there is a limited number of well-characterized compounds containing (2) two different nucleobases bound to a single metal,[45,46] (3) two different metals bound to a single nucleobase,[47,48] and (4) two different metals simultaneously bound to two different nucleobases.[45,49] The four situations are schematically depicted in

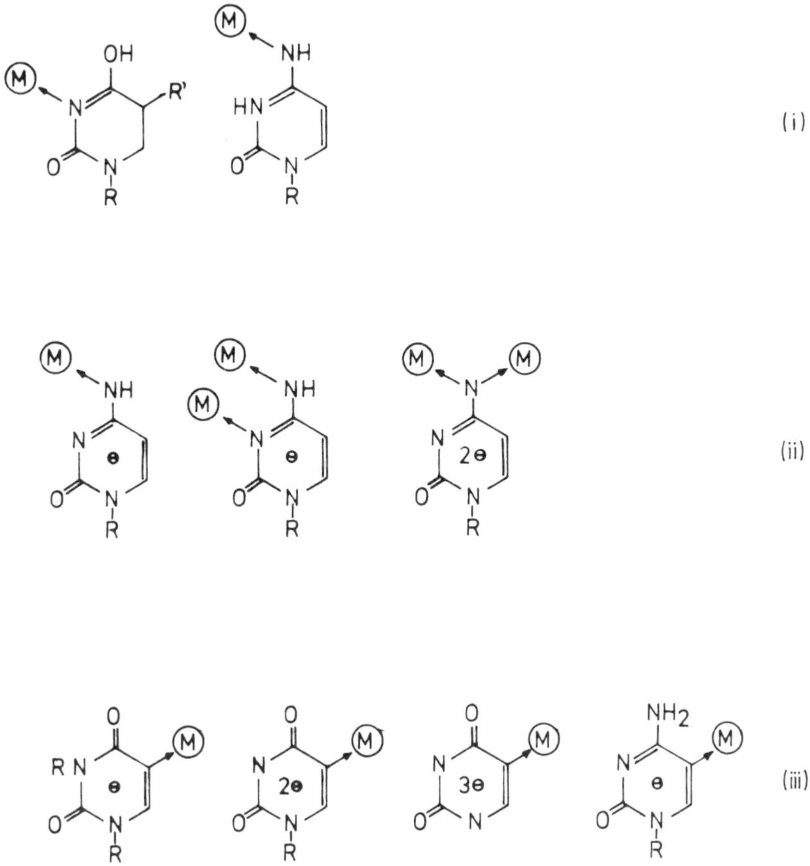

FIGURE 3. (i) Metal-stabilized rare tautomer of uracil (thymine) and cytosine. (ii) Unusual anions of N1-substituted cytosine in metal-complexed form. (iii) Carbon-bound uracil and cytosine nucleobases leading to rare anionic forms of these bases.

Figure 4. Examples for (2) to (4) almost exclusively contain Pt(II) as M_1 (compared to few Hg(II) species) and at least one 1-methyluracil or 1-methylthymine anion (L_1). It is the specific distribution of negative charge in N3-metalated uracil (and thymine) which facilitates binding of additional metal electrophiles via the exocyclic oxygens.

SCOPE

Reviews dealing with metal complexes of uracil and thymine,[50] as well as of Pt complexes of uracil, thymine, and cytosine[51] have appeared. Crystal structures of metal complexes of these bases were reviewed in 1979,[8] and of Pt complexes in 1981.[10] Data included in the present review are from structural, spectroscopic,

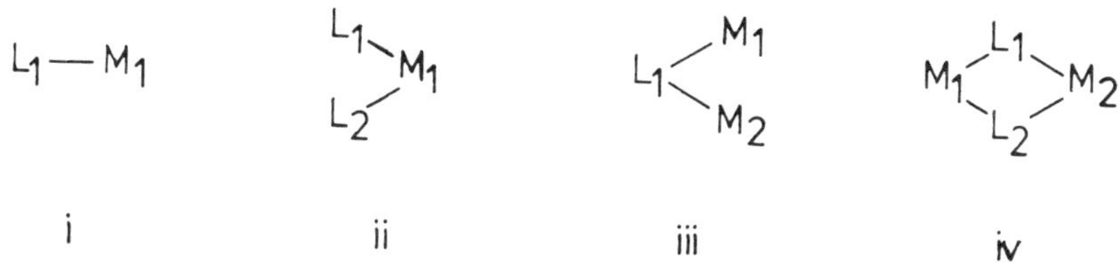

FIGURE 4. Schematic representations of metal-nucleobase complexes comprised of (i) single metal M_1, single nucleobase L_1, (ii) single metal M_1, different nucleobases L_1 and L_2, (iii) two different metals (M_1, M_2), single nucleobase L_1, and (iv) two different metals (M_1, M_2) and different nucleobases (L_1, L_2).

physicochemical, as well as theoretical studies. Unfortunately, not all of the authors' conclusions reported hereafter appear to be substantiated by experimental findings, quite a few are speculative or even arbitrary and are for this reason of limited value to the reader. This applies in particular to complexes containing the monoanions of unsubstituted uracil and thymine ligands, when the possibility of tautomerism and linkage isomerism is completely ignored. Similarly, multiple metal binding to these bases is only rarely considered, although it now begins to emerge as an outstanding property of uracil and thymine coordination chemistry. The reader is therefore advised to make use of the reference with proper care.

The pyrimidine nucleobases considered in this review include, first, the unsubstituted nucleobases uracil, thymine, and cytosine, second, their N1-alkylated derivatives, which are models of the respective nucleosides and nucleotides, and, third, selected uncommon or pharmacologically interesting pyrimidine nucleobases, as well as orotic acid as a precursor of the natural pyrimidine bases. The tables list metal compounds of the parent ligands cytosine, uracil, and thymine, as well as the N1-methylated derivatives, followed by derivatives which are listed with increasing degree of substitution. Charges of ligands are omitted. In most cases they can be deduced from the given stoichiometry and the metal oxidation state. Similarly, binding sites are not indicated and differentiated, respectively.

As far as metals are concerned, compounds containing exclusively main group metals (e.g., alkali salts) are excluded, as are salts containing no directly coordinated transition metal[52] or compounds where the interaction between a transition metal complex and a nucleobase is through hydrogen bonding only.[53] Water of crystallization is not listed for the metal compounds unless determined by X-ray crystallography.

TABLE INDEX

Base no.	Compound	Base
C1	Cytosine	cyt
C2	1-Methylcytosine	1mecyt
C3	6-Hydroxy-2-thiocytosine	6HO2Scyt
C4	6-Amino-2-thiocytosine	6A2Scyt
C5	5-Amino-6-hydroxycytosine	5A6HOcyt
C6	5,6-Diamino-2-thiocytosine	56dA2Scyt
C7	5-Amino-6-hydroxy-2-thiocytosine	5A6HO2Scyt
U3	Thymine	see uracil
U1	Uracil	ura
U2	1-Methyluracil	1meura
U3	Thymine	thy
U4	1-Methylthymine	1methy
U5	2-Thiouracil	2Sura
U6	*S,S*-2-Thiouracil	2SSura
U7	4-Thiouracil	4Sura
U8	5-Bromouracil	5Brura
U9	5-Chlorouracil	5Clura
U10	5-Fluorouracil	5Fura
U11	5-Iodouracil	5Iura
U12	5-Nitrouracil	$5NO_2ura$
U13	5-Ruracil	5Rura
U14	6-Methyluracil	6meura
U15	6-Azauracil	6azura
U16	Orotic acid	oro
U17	1,3-Dimethyluracil	13dmeura
U18	5-Chloro-1-methyluracil	5Cl1meura
U19	1-Methyl-5-nitrouracil	$1me5NO_2ura$
U20	2,4-Dithiouracil	24dSura
U21	6-Methyl-2-thiouracil	6me2Sura
U22	6-Amino-2-thiouracil	6A2Sura
U23	3-Methylorotic acid	3meoro
U24	5,6-Dihydrouracil	56dHura
U25	5,6-Diaminouracil	56dAura
U26	5-Nitroorotic acid	$5NO_2oro$
U27	5,6,-Dihydro-5,6-dihydroxythymine	56dHdHOthy
U28	6-Amino-5-nitrouracil	$6A5NO_2ura$
U29	6-Amino-5-formyluracil	6A5fura
U30	Dithymine	dthy
U31	Diorotic acid	doro
U32	5,6-Dihydro-1-methyluracil	56dH1meura
U33	5,5-Dichloro-6-hydroxy-5,6-dihydro-1-methyluracil	55dCl6HO56dH1-meura
U34	1,3-Dialkyl-5-fluorouracil	13dme5Fura 13det5Fura 13dpr5Fura
U35	6-Amino-1-methyl-5-nitrosouracil	6A1me5NOura

TABLE INDEX (CONTINUED)

Base no.	Compound	Base
U36	6-Amino-3-methyl-5-nitrosouracil	6A3me5NOura
U37	6-Amino-1,3-dimethyl-5-nitrosouracil	6A13dme5NOura
U38	6-Amino-5-formyl-1-methyluracil	6A5f1meura
U39	6-Amino-5-formyl-3-methyluracil	6A5f3meura
U40	6-Amino-1,3-dimethyl-5-formyluracil	6A13dme5fura
U41	5,6-Dihydro-5,6-dihydroxy-1-methylthymine	56dHdHO1methy

TABLE 1

Base no.	Base	Metal	Stoichiometry	Method	Ref.
			Cytosine		
C1	cyt	Au(I)	$[R_2SAu(cyt)]Cl$	nmr	140
		Au(III)	$AuCl_3(cyt)$	nmr	140
		Ce(III)	$CeCl_3(cyt)_2$	uv, pot, cond	141
		Co(II)	$[Co(cyt)_2](SCN)_2$		142
			$[Co(cyt)_2]^{2+}$	pot	143
			$[Co(cyt)]^{2+}$	pot	7
			$[Co(cyt)(his)]^{n+}$	pot	144
			$[Co(cyt)(trp)]^{n+}$	pot	144
			$[Co(cyt)(ade)]^{n+}$	pot	10, 143
			$[Co(cyt)(phen)]^{n+}$	pot	10, 143
			$[Co(cyt)(bipy)]^{n+}$	pot	10, 143
			$[Co(cyt)(5'\text{-}SSA)]^{n+}$	pot	10, 143
			$[Co(cyt)(gly)]^{n+}$	pot	145
			$[Co(cyt)(ox)]^{n+}$	pot	145
			$[Co(cyt)(hisam)]^{n+}$	pot	145
		Cu(II)	$[Cu(cyt)]^{2+}$	pot	7
				calc	13
			$[Cu(cyt)(his)]^{n+}$	pot	144
			$[Cu(cyt)(trp)]^{n+}$	pot	144
			$[Cu(cyt)(ade)]^{n+}$	pot	10, 143
			$[Cu(cyt)(phen)]^{n+}$	pot	10, 143
			$[Cu(cyt)(bipy)]^{n+}$	pot	10, 143
			$[Cu(cyt)(5'\text{-}SSA)]^{n+}$	pot	10, 143
			$[Cu(cyt)(gly)]^{n+}$	pot	145
			$[Cu(cyt)(ox)]^{n+}$	pot	145
			$[Cu(cyt)_4](ClO_4)_2$		142
			$[Cu(cyt)_2]^{2+}$	pot	143
			$Cu(cyt)_2Cl_2$	X-ray	146
			$Cu(glygly)(cyt).2H_2O$	X-ray	147, 148
			$[Cu(salen)(cyt)]NO_3.H_2O$	X-ray	149
			$[Cu(cyt)_4](C1O_4)_2$	ir	150
			$Cu_2(cyt)_4(EtOH)(ClO_4)_3$	ir	150
		Ir(I)	*trans*-$Ir(CO)(PPh_3)_2(cyt)$	ir	18
		Ir(III)	$IrCl_3(cyt)_2(MeOH)$	ir	151
		Mn(II)	$[Mn(cyt)(gly)]^{n+}$	pot	145
			$[Mn(cyt)(ox)]^{n+}$	pot	145
			$[Mn(cyt)(hisam)]^{n+}$	pot	145
		Ni(II)	$[Ni(cyt)_3(H_2O)_3](SCN)_2$		142
			$[Ni(cyt)_4(H_2O)_2](SCN)_2$		142
			$[Ni(cyt)]^{2+}$	pot	7
			$[Ni(cyt)_2]^{2+}$	pot	143
			$[Ni(cyt)(his)]^{n+}$	pot	144
			$[Ni(cyt)(trp)]^{n+}$	pot	144
			$[Ni(cyt)(ade)]^{n+}$	pot	9, 143

TABLE 1 (CONTINUED)

Base no.	Base	Metal	Stoichiometry	Method	Ref.
			$[Ni(cyt)(phen)]^{n+}$	pot	9, 143
			$[Ni(cyt)(bipy)]^{n+}$	pot	9, 143
			$[Ni(cyt)(5'\text{-}SSA)]^{n+}$	pot	9, 143
			$[Ni(cyt)(gly)]^{n+}$	pot	145
			$[Ni(cyt)(ox)]^{n+}$	pot	145
			$[Ni(cyt)(hisam)]^{n+}$	pot	145
		Pd(II)	$PdCl_2/(cyt)$	nmr	152
			trans-$Pd(cyt)_2Cl_2$	ir	153
			trans-$Pd(cyt)_2Br_2$	ir	153
		Pt(II)	$PtCl_2/(cyt)$	nmr	152
			$[PtCl_3(cyt)]^+$	X-ray, nmr, ram	154
			cis-$PtCl_2(cyt)(1mecyt)$		154
			cis-$PtCl_2(cyt)(DMSO)$		154
			$[(NH_3)_3Pt(cyt)]Cl_2$		154
			$[(NH_3)_3Pt(cyt)Pt(dien)]^{3+}$	nmr	154
			$Pt(cyt)I_2$	ir, nmr	155
		Pt(n) (n > 2)	Pt-(cyt)-blue	esr	69
		Rh(I)	*trans*-$Rh(PPh_3)_2(CO)(cyt)$	ir	18
			cis-$Rh(CO)_2Cl(cyt)$	nmr	156
		Rh(III)	$RhCl_3(cyt)_2(MeOH)_2$	ir	151
		Ru(II)	$Ru_2Cl_4(cyt)_4(DMSO)_2(MeOH)_4$	ir	151
		Ru(III)	$RuCl_3(cyt)(MeOH)_2$	ir	151
		Th(IV)	$Th(cyt)_2(NO_3)_4$	ir, uv, cond	141
		U(VI)	$UO_2Cl_2(cyt)_2$	ir, uv, cond	141
		Zn(II)	$[Zn(cyt)]^{2+}$	pot	7
			$[Zn(cyt)(his)]^{n+}$	pot	144
			$[Zn(cyt)(trp)]^{n+}$	pot	144
			$[Zn(cyt)(ade)]^{n+}$	pot	10, 143
			$[Zn(cyt)(phen)]^{n+}$	pot	10, 143
			$[Zn(cyt)(bipy)]^{n+}$	pot	10, 143
			$[Zn(cyt)(5'\text{-}SSA)]^{n+}$	pot	10, 143
			$[Zn(cyt)(gly)]^{n+}$	pot	145
			$[Zn(cyt)(ox)]^{n+}$	pot	145
			$[Zn(cyt)(hisam)]^{n+}$	pot	145
			$[Zn(cyt)_2]^{2+}$	pot	143
			1-Methylcytosine		
C2	1mecyt	Ag(I)	$[Ag(1mecyt)(NO_3)]_2$	X-ray	157
		Au(III)	$AuCl_3(1mecyt)$	X-ray	158
		Cd(II)	$CdCl_2(1mecyt)_2$	X-ray	159
		Hg(II)	$[HgCl_2(1mecyt)]_2$	X-ray	160
			$[(MeHg)(1mecyt)]NO_3$	nmr	161
			$[(MeHg)_2(1mecyt)]NO_3$	X-ray, ir, nmr	161
			$[(MeHg)_2(1mecyt)]ClO_4$		161
			$[(MeHg)_3(1mecyt)]NO_3$	X-ray	162

TABLE 1 (CONTINUED)

Base no.	Base	Metal	Stoichiometry	Method	Ref.
		Pd(II)	*trans*-$PdCl_2(1mecyt)_2$	ir	153
				X-ray	163
		Pt(II)	*cis*-$[Pt(NH_3)_2Cl(1mecyt)]NO_3$	X-ray, ir, ram, nmr	164
			cis-$[Pt(NH_3)_2Cl(1mecyt)]_2[Pt(CN)_4]$	X-ray	165
			cis-$[Pt(NH_3)_2(H_2O)(1mecyt)](NO_3)_2 \cdot H_2O$	X-ray	166
			cis-$[Pt(NH_3)_2(NO_3)(1mecyt)]NO_3$	ir	166
			cis-$[Pt(NH_3)_2(OH)(1mecyt)]NO_3$		166
			cis-$[Pt(NH_3)_2(OH)(1mecyt)]NO_3 \cdot 2H_2O$	X-ray	166
			cis-$[Pt(NH_3)_2(thy)(1mecyt)]COO_4$	ir, ram, X-ray	68
			cis-$[Pt(NH_3)_2(thy)(1mecyt)]ClO_4 \cdot H_2O$	ir, ram, X-ray	68
			cis-$[Pt(NH_3)_2(thy)(1mecyt)]ClO_4 \cdot 3H_2O$	X-ray	66
				ir, ram	68
			cis-$[Pt(NH_3)_2(9meade)(1mecyt)](ClO_4)_2.H_2O$	X-ray, nmr	167
			cis-$[Pt(NH_3)_2(9meade)(1mecyt)](ClO_4)_3$		167
			cis-$[Pt(NH_3)_2(9meade)(1mecyt)]^{2+}$	nmr	167
			cis-$[Pt(NH_3)_2(1mecyt)(9meade)Pt(1mecyt)(NH_3)_2]^{4+}$	nmr	167
			cis-$[Pt(NH_3)_2(9etgua)(1mecyt)](ClO_4)_2$	X-ray, ir, ram	168
				nmr	169
			cis-$[Pt(NH_3)_2(9etgua)(1mecyt)]ClO_4 \cdot 4H_2O$	X-ray, ir, ram	168
				nmr	169
			cis-$[Pt(NH_3)_2(9etgua)(1mecyt)][Pt(NH_3)_2(9etgua)(1mecyt)](ClO_4)_3$	X-ray	168, 170
				ir, ram	168
				nmr	169
			cis-$[Pt(NH_3)_2(1meura)(1mecyt)]NO_3$	nmr	38

TABLE 1 (CONTINUED)

Base no.	Base	Metal	Stoichiometry	Method	Ref.
			cis-$[Pt(NH_3)_2(1mecyt)_2]Cl_2$		171
			cis-$[Pt(NH_3)_2(1mecyt)_2](NO_3)_2 \cdot (1mecyt)$	X-ray, nmr, ir, ram	171, 172
			cis-$[Pt(NH_3)_2(1mecyt)_2][Pt(CN)_4] \cdot 2H_2O$	X-ray	165
			trans-$[Pt(NH_3)_2(9meade)(1mecyt)](ClO_4)_2$	X-ray, nmr	167
			trans-$[Pt(NH_3)_2(9meade)(1mecyt)]^{2+}$	nmr	167
			trans-$[Pt(NH_3)_2(1mecyt)(9meade)Pt(NH_3)_2(1mecyt)]^{4+}$	nmr	172
			trans-$[Pt(NH_3)_2(9etgua)(1mecyt)](ClO_4)_2$	X-ray	173
			trans-$PtCl_2(NH_3)(1mecyt)OO \cdot 5 H_2O$	X-ray	174
				ms	175
			trans-$PtCl_2(DMSO)(1mecyt)$	X-ray	176
			trans-$[Pt(NH_3)_2(1mecyt)_2](OO_3)_2$	X-ray	174
			trans-$[Pt(NH_3)_2(1mecyt)_2](ClO_4)_2$	X-ray	177
			$[Pt(NH_3)(1mecyt)_3](ClO_4)_2 \cdot H_2O$	X-ray	178
				nmr	179
			trans-$[Pt(9etgua)_2(NH_3)(1mecyt)](ClO_4)_2$	nmr	179
			cis-$[(NH_3)_2Pt(1mecyt)_2PtNH_3)_2](NO_3)_2 \cdot 2H_2O$	X-ray, nmr	180
		Pt(III)	*cis*-$[(NO_2)(NH_3)_2Pt(1mecyt)_2Pt(NH_3)_2(NO_2)](NO_3)_2 \cdot 2H_2O$	X-ray	180
		Pt(IV)	*cis,trans*-$Pt(NH_3)_2(OH)_2Cl(1mecyt)$	X-ray	181
			trans,trans,trans-$[Pt(NH_3)_2(OH)_2(1mecyt)_2](NO_3)_2 \cdot 2H_2O$	X-ray	182
			trans,trans-$[Pt(NH_3)_2(1mecyt)_2(H_2O)(OH)](NO_3)_3 \cdot 3H_2O$	X-ray	182
			trans,trans,trans-$[Pt(NH_3)_2(OH)_2(1mecyt)_2](NO_3)_2 \cdot 2H_2O$	X-ray, ram	183

TABLE 1 (CONTINUED)

Base no.	Base	Metal	Stoichiometry	Method	Ref.
			trans-$[Pt(NH_3)_2(1mecyt)_2(OH)](NO_3)_2 \cdot H_2O$	X-ray	182
			trans,trans-$[Pt(NH_3)_2(1mecyt)_2](NO_3)_2 \cdot 2H_2O$	X-ray	182, 184
		Pt(II)/ Ag(I)	*cis*-$[Pt(NH_3)_2(1meura)(1mecyt)Ag(H_2O)](NO_3)_2 \cdot AgNO_3 \cdot 2{\cdot}5H_2O$	X-ray	54
		Pt(II)/ Cu(II)	*cis*-$[Pt(NH_3)_2(1meura)(1mecyt)Cu(1mecyt)(1meura)Pt(NH_3)_2](NO_3)_4 \cdot 6H_2O$	X-ray, esr	38
		Rh(I)	*cis*-$Rh(CO)_2Cl(1mecyt)$	ir	156
		Ru(III)	$[Ru(NH_3)_5(1mecyt)](PF_6)_2$	X-ray	185
		Zn(II)	$ZnCl_2(1mecyt)_2$	X-ray	186
			6-Hydroxy-2-thiocytosine		
C3	6H02 Scyt	Co(II)	$Co(6HO2Scyt)_2(H_2O)_2$		187
		Ni(II)	$[Ni(6HO2Scyt)(H_2O)_2](NO_3)_2$		187
			$Ni(6HO2Scyt)_2$		187
			$Ni(6HO2Scyt)_2(H_2O)_2$		187
			6-Amino-2-thiocytosine		
C4	6A2S cyt	Co(II)	$[Co(6A2Scyt)_2]Cl_2$		187
			$[Co(6A2Scyt)_2(H_2O)_2](NO_3)_2$		187
			$[Co(6A2Scyt)_2](SCN)_2$		187
			$[Co(6A2Scyt)_2(H_2O)_2]SO_4$		187
		Cu(I)	$[Cu_3Cl_2(6A2Scyt)_2]Cl \cdot 2HCl$		90
		Ni(II)	$[Ni(6A2Scyt)_3]SO_4$		187
			$[Ni(6A2Scyt)_2]Br_2$		187
			$[Ni(6A2Scyt)_2](NO_3)_2$		187
			$Ni(6A2Scyt)_2(SCN)_2$		187
			$[Ni(6A2Scyt)_2(H_2O)_2](ClO_4)_2$		187
			$Ni(6A2Scyt)_2$		187
			5-Amino-6-hydroxycytosine		
C5	5A6H-Ocyt	Cu(II)	$[Cu(5A6HOcyt)]SO_4$		127
			5,6-Diamino-2-thiocytosine		
C6	56dA-2Scyt	Cu(I)	Cu(56dA2Scyt)Cl.1.5HCl		127

TABLE 1 (CONTINUED)

Base no.	Base	Metal	Stoichiometry	Method	Ref.
		Ni(II)	[Ni(56dA2Scyt)]SO_4		127
			[Ni(56dA2Scyt)$(H_2O)_2$]SO_4	mag	127
			5-Amino-6-hydroxy-2-thiocytosine		
C7	5A6HO-2Scyt	Co(II)	[Co$(5A6HO2Scyt)_3$]SO_4		127
			Uracil		
U1	ura	Ag(I)	Ag(ura)	pot	1
				nmr, uv	2
			$[Ag(ura)_2]^-$	nmr, uv	2
			$[Ag(ura)]^-$	nmr, uv	2
			$[Ag(ura)_2]^{3-}$	nmr, uv	2
		Cd(II)	$CdCl_2$(ura)	ir, uv	3
			$Cd(ura)_2(H_2O)_3$	X-ray	4
		Co(II)	$CoCl_2$(ura)	ir, uv	3
			$[Co(ura)_4(H_2O)_2]Cl_2$	ir, uv	5
			$Na[Co(gly)_2(ura)_2]$	ir, mag, uv	6
			$[Co(ura)]^+$	pot	7
			$Co(ura)_2(H_2O)_2$	ir, uv	8
			$[Co(ura)(bipy)]^+$	pot, thermodyn	9, 10
			$[Co(ura)(phen)]^+$	pot, thermodyn	9, 10
			$[Co(ura)(ade)]^+$	pot, thermodyn	9, 10
			$Co(ura)_2$	ir, uv	9
			$[Co(ura)(asp)]^-$	pot	11
			$[Co(ura)(glu)]^-$	pot	12
		Cu(II)	$CuCl_2$(ura)	ir, uv	3
			$[Cu(ura)_4(H_2O)_2]SO_4$	ir, uv	5
			$Na_2[Cu(gly)(ura)(OH)_2]$	ir, cond, mag	6
			Cu(glygly)(ura).HBr	ir	14
			$[Cu(ura)(asp)]^-$	pot	11
			$[Cu(ura)(glu)]^-$	pot	12
			$[Cu(ura)]^+$	pot	1, 7
				calc	13
			$Cu(ura)_2(H_2O)_2$	ir, uv	8
			$[Cu(ura)(ade)]^+$	pot, thermodyn	9, 10
			$[Cu(ura)(phen)]^+$	pot, thermodyn	9, 10
			$[Cu(ura)(bipy)]^+$	pot, thermodyn	9, 10
			$Cu(ura)_2$	pot	9
		Fe(II)	$Fe(ura)_2(H_2O)_2$	ir, uv	8

TABLE 1 (CONTINUED)

Base no.	Base	Metal	Stoichiometry	Method	Ref.
		Hg(II)	$HgCl_2(ura)$	X-ray	15
				ir, ram	16
			$[Hg(ura)]^+$	pot, thermodyn	17
			$Hg(ura)_2$	pot, thermodyn	17
		Ir(I)	*trans*-$Ir(CO)(PPh_3)_2(ura)$	ir	18
		La(III)	$La(ura)_2$	ir	19
		Pr(III)	$Pr_2(ura)_3$	ir	19
		Nd(III)	$Nd_2(ura)_3$	ir	19
		Mn(II)	$MnCl_2(ura)$	ir, uv	3
			$MnCl_2(ura)_2$	ir, uv	5
			$[Mn(ura)]^+$	pot	7
			$[Mn(ura)(ade)]^+$	pot, thermodyn	9, 10
			$[Mn(ura)(phen)]^+$	pot, thermodyn	9, 10
			$[Mn(ura)(bipy)]^+$	pot, thermodyn	9, 10
			$Mn(ura)_2$	pot	9
			$Mn(ura)_2(H_2O)_2$	ir, uv	8
		Ni(II)	$NiCl_2(ura)$	ir, uv	3
			$NiCl_2(ura)_2$	ir, uv	5
			$[Ni(AcO)_2(ura).H_2O]_2$	ir, cond, spec	20
			$[Ni(ura)(asp)]^-$	pot	11
			$[Ni(ura)(glu)]^-$	pot	12
			Ni(gly)(ura)	ir, cond, mag	6
			$[Ni(ura)]^+$	pot	1, 7
			$[Ni(ura)(ade)]^+$	pot, thermodyn	9, 10
			$[Ni(ura)(phen)]^+$	pot, thermodyn	9, 10
			$[Ni(ura)(bipy)]^+$	pot, thermodyn	9, 10
			$Ni(ura)_2$	pot	9
			$Ni(ura)_2(H_2O)_2$	ir, uv	8
			$Ni(NH_3)_2(ura)_2$		21
			trans, trans, trans-$Ni(NH_3)_2(ura)_2(H_2O)_2$	X-ray	22
		Pd(II)	$[Pd(dien)(ura)]^+$	pot	23
			$[Pd(en)(ura)(H_2O)]^+$	pot	23
			$Pd(en)(ura)_2$	pot	23
		Pt(II)	$[enPt(ura)Cl]Cl.2H_2O$	X-ray	24
			$[(NH_3)_3Pt(ura)]NO_3$	ir, ram, nmr	25, 26
			$[(NH_3)_3Pt(ura)]Cl$	ir, ram, nmr	25, 26
			enPt(ura)Cl.DMF	ir, ram, nmr	26
			enPt(ura)Cl	ir, ram, nmr	26

TABLE 1 (CONTINUED)

Base no.	Base	Metal	Stoichiometry	Method	Ref.
			$enPt(ura)_2$	ir, ram, nmr	26
			cis-$(NH_3)_2Pt(ura)Cl$	ir, ram, nmr	26
			cis-$(NH_3)_2Pt(ura)_2$	ir, ram, nmr	26
			$[enPt(ura)]_n(NO_3)_n$	ir, ram, nmr	26
			$[en_8Pt_8(ura)_4](NO_3)_8.mH_2O$	X-ray	27
			$(NH_3)_3Pt(ura)$	ir, ram, nmr	26
			$Na_2[(NH_3)_2Pt(ura)_2]$	ir, ram, nmr	26
		Pt(n)	Pt-uracil blues	esr	28, 29
		(n > 2)		prep	30
			Pt(ura)...DNA		31
		Rh(I)	*trans*-$Rh(CO)(PPh_3)_2(ura)$	ir	16
		Ti(II)	$(MeCp)_2Ti_2(ura)$	esr, mag	32
		Zn(II)	$Zn(ura)_2(ClO_4)_2$	ir, uv	3
			$Na_2[Zn(gly)(ura)(OH)_2(H_2O)]$	ir	6
			$[Zn(ura)]^+$	pot	7
			$[Zn(ura)(ade)]^+$	pot, thermodyn	9, 10
			$[Zn(ura)(phen)]^+$	pot, thermodyn	9, 10
			$[Zn(ura)(bipy)]^+$	pot, thermodyn	9, 10
			$Zn(ura)_2$	pot	9
			$[Zn(ura)(asp)]^-$	pot	11
			$[Zn(ura)(glu)]^-$	pot	12
			1-Methyluracil		
U2	1meura	Ag(I)	$[Ag(1meura)]_n$	X-ray	33
		Pt(II)	*cis*-$(NH_3)_2Pt(1meura)Cl$		34, 37
			cis-$(NH_3)_2Pt(1meura)I$		35
			cis-$[(NH_3)_2Pt(1meura)(H_2O)]NO_3$	nmr	34
			cis-$(NH_3)_2Pt(1meura)_2.4H_2O$	X-ray	36
				nmr, ram	34
			cis-$[(NH_3)_2(1meura)Pt(OH)Pt(1meura)(NH_3)_2]ClO_4$	nmr	34
			cis-$[(NH_3)_2Pt(1meura)(1mecyt)]NO_3$	nmr	38
			cis-$[(NH_3)_2Pt(1meura)_2Pt(NH_3)_2](NO_3)_2.H_2O$	X-ray	34
			cis-$[(NH_3)_2Pt(1meura)_2Pt(en)](NO_3)_2$	ir, nmr	39
			cis-$[(NH_3)_2Pt(1meura)_2Pt(bipy)](NO_3)_2$	ir, nmr	39

TABLE 1 (CONTINUED)

Base no.	Base	Metal	Stoichiometry	Method	Ref.
			cis-[(en)Pt(1meura)$_2$Pt(NH$_3$)$_2$](ClO$_4$)$_2$	ir, nmr	39
			[(en)Pt(1meura)$_2$Pt(en)](NO$_3$)$_2$	ir, nmr	39
			[(bipy)Pt(1meura)$_2$Pt(bipy)](NO$_3$)$_2$	ir, nmr	39
			[(bipy)Pt(1meura)$_2$Pt(bipy)](ClO$_4$)$_2$	ir, nmr	39
			cis-[(NH$_3$)$_2$Pt(1meura)$_2$Pt(NH$_3$)$_2$](NO$_3$)$_2$.H$_2$O	X-ray	40
			cis-[(NH$_3$)$_2$Pt(1meura)$_2$Pt(NH$_3$)$_2$]Cl$_2$	ir, ram	35
			cis-[(NH$_3$)$_2$Pt(1meura)$_2$Pt(NH$_3$)$_2$]I$_{1.5}$(NO$_3$)$_{0.5}$	ir	35
			cis-(NH$_3$)$_2$(1meura)Pt(NC)Pt(CN)$_2$(CN)Pt(1meura) (NH$_3$)$_2$	ir	41
		Pt (2.08)	Pt-(1meura)-blue	cv, titr	42
		Pt (2.25)	[(NH$_3$)$_8$Pt$_4$(1meura)$_4$](NO$_3$)$_5$·5H$_2$O	X-ray spec, titr, esr	43 39, 44, 53
		Pt(III)	[(H$_2$O)(NH$_3$)$_2$Pt(1meura)$_2$Pt(NH$_3$)$_2$(NO$_2$)](NO$_3$)$_3$.5H$_2$O	X-ray	45
			[(H$_2$O)(NH$_3$)$_2$Pt(1meura)$_2$Pt(NH$_3$)$_2$(ONO$_2$)](NO$_3$)$_3$.3H$_2$O	X-ray	35
			[(H$_2$O)(NH$_3$)$_2$Pt(1meura)$_2$Pt(NH$_3$)$_2$(ONO$_2$)](NO$_3$)$_3$.2H$_2$O	X-ray	35
			[(NO$_2$)(NH$_3$)$_2$Pt(1meura)$_2$Pt(NH$_3$)$_2$(NO$_2$)](NO$_3$)$_2$	ir, ram	35,45
			[(NO$_2$)(NH$_3$)$_2$Pt(1meura)$_2$Pt(NH$_3$)$_2$(OH$_2$](NO$_3$)$_3$	ir, ram	35
			[(NO$_2$)(NH$_3$)$_2$Pt(1meura)$_2$Pt(NH$_3$)$_2$Cl](NO$_3$)$_{0.5}$Cl$_{1.5}$	ir, ram	35
			[Cl(NH$_3$)$_2$Pt(1meura)$_2$Pt(NH$_3$)$_2$Cl](NO$_3$)$_{1.75}$Cl$_{0.25}$	ir, ram	35

TABLE 1 (CONTINUED)

Base no.	Base	Metal	Stoichiometry	Method	Ref.
			$[(NO_2)(NH_3)_2Pt(1meura)_2Pt(NH_3)_2](NO_3)_3.H_2O$	X-ray	46
			$[Cl(NH_3)_2Pt(1meura)_2Pt(NH_3)_2Cl]Cl_2.3.5H_2O$	X-ray	47
			$Cl(NH_3)_2Pt(1meura)_2PtCl_3.2H_2O$	X-ray	47
			$[(1meura)(NH_3)_2Pt(1meura)_2Pt(NH_3)_2](NO_3)_3.2HNO_3$		48
			$[(1meura)(NH_3)_2Pt(1meura)_2Pt(NH_3)_2](NO_3)_3\cdot 7H_2O$		48
			$[(1meura)(NH_3)_2Pt(1meura)_2Pt(NH_3)_2](SiF_6)(NO_3).7H_2O$	X-ray	48
		Pt(IV)	*cis*-$(NH_3)_2Pt(1meura)Cl_3$	nmr	49
		Pt(II)/Ag(I)	*trans*-$(NH_3)_2Pt(1meura)_2Ag_2(NO_3)_2(H_2O).H_2O$	X-ray	50
			cis-$[(NH_3)_2Pt(1meura)_2Ag(1meura)_2Pt(NH_3)_2]NO_3$		51
			cis-$(NH_3)_2Pt(1meura)_2.AgNO_3$		51
			cis-$(NH_3)_2Pt(1meura)_2.1.5AgClO_4$		51
			cis-$[(NH_3)_2(ONO_2)Pt(1meura)Ag]NO_3$	ir, ram	52
			cis-$[(NH_3)_4Pt_2(1meura)_2Ag_2](NO_3)_4.2H_2O$	X-ray	52
			cis-$[(NH_3)_8Pt_4(1meura)_4Ag](NO_3)_5.4H_2O$	X-ray	44
			cis-$[(NH_3)_4Pt_2(1meura)_2Ag](NO_3)_3.AgNO_3.0.5H_2O$	X-ray	53
			cis-$[(NH_3)_2Pt(1meura)(1mecyt)Ag(H_2O)](NO_3)_2.AgNO_3.2.5H_2O$	X-ray	54
		Pt(II)/Co(II)	*cis*-$[(NH_3)_2Pt(1meura)_2Co(1meura)_2Pt(NH_3)_2](NO_3)_2$	uv	51
				spec	55
				mag	56
			cis-$[(NH_3)_2Pt(1meura)_2Co(1meura)_2Pt(NH_3)_2](ClO_4)_2$	uv	51

TABLE 1 (CONTINUED)

Base no.	Base	Metal	Stoichiometry	Method	Ref.
		Pt(II)/ Cu(II)	*cis*-$[(NH_3)_2Pt(1meura)_2Cu(H_2O)_2]$ $SO_4.4.5H_2O$	X-ray	36
			cis-$[(NH_3)_2Pt(1meura)_2Cu$ $(1meura)_2Pt(NH_3)_2]SO_4.(H_2O)_x$	X-ray	57
				spec	55
			cis-$[(NH_3)_2Pt(1meura)_2Cu$ $(1meura)_2Pt(NH_3)_2](NO_3)_2$		36
			cis-$[(NH_3)_2Pt(1meura)_2Cu$ $(1meura)_2Pt(NH_3)_2]Cl_2$		36
			cis-$[(NH_3)_2Pt(1meura)_2Cu$ $(1meura)_2Pt(NH_3)_2][PtCl_4]$	ir	36
			cis-$[(NH_3)_2Pt(1mecyt)$ $(1meura)Cu(1meura)(1mecyt)$	X-ray, esr	38
			$Pt(NH_3)_2](NO_3)_4.6H_2O$		
		Pt(II)/ Fe(II)	*cis*-$(NH_3)_2Pt(1meura)_2.FeSO_4$	ir	51
			cis-$[(NH_3)_2Pt(1meura)_2Fe$ $(1meura)_2Pt(NH_3)_2](NO_3)_2$	spec, MB	51
		Pt(II)/ Fe(III)	*cis*-$[(NH_3)_2Pt(1meura)_2Fe$ $(1meura)_2Pt(NH_3)_2](NO_3)_3$		51
		Pt(II)/ Ni(II)	*cis*-$[(NH_3)_2Pt(1meura)_2Ni$ $(1meura)_2Pt(NH_3)_2](NO_3)_2$		51
				spec	55
				mag	56
		Pt(II)/ Pd(II)	*cis*-$(NH_3)_2Pt(1meura)_2PdCl_2$		58b
			cis-$[(NH_3)_2Pt(1meura)_2Pd(en)]$ $(NO_3)_2.6H_2O$	X-ray, nmr	39
			cis-$[(NH_3)_2Pt(1meura)_2Pd(bipy)]$ $(NO_3)_2$	ir	39
			$[(en)Pt(1meura)_2Pd(en)]SO_4$	ir, nmr	39
			$[(en)Pt(1meura)_2Pd(bipy)](NO_3)_2$	ir, nmr	39
			$[(bipy)Pt(1meura)_2Pd(bipy)]$ $(NO_3)_2$	ir	39
			cis-$[(NH_3)_2Pt(1meura)_2Pd$ $(1meura)_2Pt(NH_3)_2]$ $(ClO_4)_2.2.25H_2O$	X-ray, cv, uv	58b

TABLE 1 (CONTINUED)

Base no.	Base	Metal	Stoichiometry	Method	Ref.
			[(en)Pt(1meura)$_2$Pd(1meura)$_2$Pt (en)](ClO_4)$_2$	ir	58
		Pt(II)/ Pd(III)	*cis*-[(NH_3)$_2$Pt(1meura)$_2$Pd (1meura)$_2$Pt(NH_3)$_2$] (NO_3)$_3$.11H_2O	X-ray, cv, uv	58
			cis-[(NH_3)$_2$Pt(1meura)$_2$Pd (1meura)$_2$Pt(NH_3)$_2$] (NO_3)$_3$.HNO_3.5H_2O	X-ray, cv, uv	58
		Pt(II)/ Zn(II)	*cis*-[(NH_3)$_2$Pt(1meura)$_2$Zn(H_2O)$_3$] SO_4.2H_2O	X-ray, nmr, ks	59
			Thymine		
U3	thy	Cd(II)	$CdCl_2$(thy)	ir, uv	3
		Co(II)	$CoCl_2$(thy)	ir, uv	3
			[Co(thy)]$^+$	pot	7, 60
			Co(thy)$_2$	pot	60
			[Co(thy)(ade)]$^+$	pot	60
			[Co(thy)(phen)]$^+$	pot	60
			[Co(thy)(bipy)]$^+$	pot	60
			[Co(thy)(asp)]$^-$	pot	11
			Co(thy)(asp)	pot	11
			[Co(thy)(glu)]$^-$	pot	12
		Cu(II)	$CuCl_2$(thy)	ir, uv	3
			[Cu(thy)]$^+$	pot	7, 60
			Cu(thy)$_2$	pot	60
			[Cu(thy)(ade)]$^+$	pot	60
			[Cu(thy)(phen)]$^+$	pot	60
			[Cu(thy)(bipy)]$^+$	pot	60
			[Cu_2(thy)$_2$]$^{2+}$	pot	61
			Cu_2(thy)$_4$	pot	61
			[Cu(dien)(H_2O)(thy)]Br.2H_2O	X-ray	62
			[Cu(thy)(asp)]$^-$	pot	11
			[Cu(thy)(glu)]$^-$	pot	12
		Hg(II)	[Hg(thy)]$^+$	pot, thermodyn	17, 61
			Hg(thy)$_2$	pot, thermodyn	17, 61
		Ir(I)	*trans*-Ir(CO)(PPh_3)$_2$(thy)	ir	18
		Mn(II)	$MnCl_2$(thy)	ir, uv	3
			[Mn(thy)]$^+$	pot	7, 60
			Mn(thy)$_2$	pot	60
			[Mn(thy)(ade)]$^+$	pot	60
			[Mn(thy)(phen)]$^+$	pot	60

TABLE 1 (CONTINUED)

Base no.	Base	Metal	Stoichiometry	Method	Ref.
			$[Mn(thy)(bipy)]^+$	pot	60
		Ni(II)	$Ni(NO_3)_2(thy)_2$	ir, uv	3
			$[Ni(thy)]^+$	pot	7, 60
			$Ni(thy)_2$	pot	60
			$[Ni(thy)(ade)]^+$	pot	60
			$[Ni(thy)(phen)]^+$	pot	60
			$[Ni(thy)(bipy)]^+$	pot	60
			$[Ni(thy)(asp)]^-$	pot	11
			Ni(thy)(asp)	pot	11
			$[Ni(thy)(glu)]^-$	pot	12
		Pt(II)	$[(NH_3)_3Pt(thy)]^{2+}$	calc	63
			$[(NH_3)_3Pt(thy)]NO_3$	nmr	64
			$[(NH_3)_3Pt(thy)](thy)$	nmr	64
			$(NH_3)_3Pt(thy)$	uv, ir, ram	64
			$[(NH_3)_3Pt(thy)]I$	nmr	64
			cis-$(NH_3)_2Pt(thy)Cl$		65
				ir	66
			trans-$(NH_3)_2Pt(thy)Cl$	ir	64
			(en)Pt(thy)Cl	X-ray	24
				ram	67
			(en)Pt(thy)Cl.2DMF	ram	67
			cis-$(NH_3)_2Pt(thy)_2$	HPLC	64, 65
			trans-$(NH_3)_2Pt(thy)_2$	HPLC	64
			trans-$[(N(CH_3)_3)_2Pt(thy)Cl]$	ram	64
			cis-$[(NH_3)_2Pt(thy)(1mecyt)]$ ClO_4	X-ray, ir, nmr, ram	68
			cis-$[(NH_3)_2Pt(thy)(1mecyt)]$ $ClO_4.3H_2O$	X-ray, ir, ram, nmr	66, 68
			trans-$[(NH_3)_2Pt(thy)(1mecyt)]$ ClO_4	ir	64
			$[(NH_3)_3Pt(thy)Pt(NH_3)_3]$ $(ClO_4)_2$	ram	64
		Pt(n) (n > 2)	Pt-(thy)-blue	esr	69
		Rh(I)	*trans*-$Rh(CO)(PPh_3)_2(thy)$	ir	18
		Zn(II)	$[Zn(thy)]^+$	pot	7, 60
			$Zn(thy)_2$	pot	60
			$[Zn(thy)(ade)]^+$	pot	60
			[Zn(thy)(phen)]	pot	60
			$[Zn(thy)(bipy)]^+$	pot	60
			$[Zn(thy)(asp)]^-$	pot	11
			$[Zn(thy)(glu)]^-$	pot	12

TABLE 1 (CONTINUED)

Base no.	Base	Metal	Stoichiometry	Method	Ref.
			1-Methylthymine		
U4	1methy	Ag(I)	$[Ag(1methy)]_n$	X-ray	70
				ir, ram	71
		Au(I)	$Au(PPh_3)(1methy)$	X-ray, nmr ir, ram	72
		Hg(II)	$Hg(1methy)_2$	X-ray	73
				ir, ram	71, 74
			$(MeHg)(1methy).0.5H_2O$	X-ray	75
				ir, ram	71
			$MeHg(1methy).0.5NaNO_3$	X-ray	75
		Pt(II)	*cis*-$(NH_3)_2Pt(1methy)Cl.H_2O$	X-ray	76
			cis-$(NH_3)_2Pt(1methy)_2$	ir	77
			cis-$[(NH_3)_2Pt(1methy)_2]Cl$	uv, ir, nmr	78
			cis-$[(NH_3)_2Pt(1methy)_2]ClO_4$	uv, ir, nmr	78
			cis-$[(NH_3)_2Pt(1methy)_2]NO_3$	uv, ir, nmr	78
			cis-$[(NH_3)_2Pt(1methy)(1mecyt)]NO_3$	nmr	79
			cis-$[(NH_3)_2Pt(1methy)(9etgua)]ClO_4$	nmr	79
			cis-$[(NH_3)_2Pt(1methy)(9meade)]ClO_4$	nmr	79
			cis-$[(NH_3)_2Pt(1methy)(9meade)](ClO_4)_2$	nmr	79
			cis-$[(NH_3)_2(1methy)Pt(9meade)Pt(1methy)(NH_3)_2](ClO_4)_2$	nmr	79
			cis-$[(NH_3)_2Pt(1methy)(9meade)]ClO_4$	nmr	79
			cis-$[(NH_3)_2Pt(1methy)_2Pt(NH_3)_2](NO_3)_2.H_2O$	X-ray	80
			cis-$[(NH_3)_2Pt(1methy)_2Pt(NH_3)_2](NO_3)_2.4.5H_2O$	X-ray, nmr	81
			cis-$[(NH_3)_2Pt(1methy)_2Pt(NH_3)_2](NO_3)_2$	X-ray	82
			cis-$[(NH_3)_2Pt(1methy)_2Pt(NH_3)Cl][Pt(NH_3)Cl_3]$	X-ray	83
			cis-$(NH_3)_2Pt(1methy)_2PtCl_2$	X-ray	83

TABLE 1 (CONTINUED)

Base no.	Base	Metal	Stoichiometry	Method	Ref.
		Pt (3.72)	Pt-(1methy)-purple	uv, mol, titr	84
		Pt(II)/ Ag(I)	*cis*-[$(NH_3)_2$Pt(1methy)$_2$Ag (1methy)$_2$Pt$(NH_3)_2$]NO_3.5H_2O	X-ray	85
			cis-$(NH_3)_2$Pt(1methy)$_2$.$AgNO_3$		85
		Pt(II)/ Co(II)	*cis*-[$(NH_3)_2$Pt(1methy)$_2$Co (1methy)$_2$Pt$(NH_3)_2$](acO)$_2$		51
		Pt(II)/ Cu(II)	*cis*-[$(NH_3)_2$Pt(1methy)$_2$Cu $(H_2O)_2$]SO_4		51
			cis-[$(NH_3)_2$Pt(1methy)$_2$Cu (1methy)$_2$Pt$(NH_3)_2$]$(NO_3)_2$		51
		Pt(II)/ Mn(II)	*cis*-[$(NH_3)_2$Pt(1methy)$_2$Mn (1methy)$_2$Pt$(NH_3)_2$]Cl_2.10H_2O	X-ray, ir, ram esr	86 87
		Pt(II)/ Ni(II)	*cis*-[$(NH_3)_2$Pt(1methy)$_2$Ni $(H_2O)_2$](acO)$_2$		51
			cis-[$(NH_3)_2$Pt(1methy)$_2$Ni (1methy)$_2$Pt$(NH_3)_2$]$(ClO_4)_2$		51
			2-Thiouracil		
U5	2Sura	Cu(I)	CuCl(2Sura)$_2$.DMF	X-ray	88
			CuCl(2Sura)		90
		Cu(II)	$Cu_2(OH)_2$(gly)(2Sura)	ir, uv, mag	6
			Na[Cu(OH)(2Sura)]		90
			Cu(2Sura)$(H_2O)_2$	ir, uv, mag	89
		Co(II)	Co(gly)(2Sura)	ir, mag	6
			Co_2(2Sura)$_5$$(H_2O)_4$	ir, uv, mag	89
		Fe(II)	Fe_2(2Sura)$_5$O	ir, mag	89
		Ni(II)	Ni(gly)(2Sura)$(H_2O)_2$	ir, mag, uv	6
			Ni(2Sura)$_2$(py)$_2$	ir, uv, mag	89
			Ni(2Sura)$_2$$(H_2O)_2$	ir, uv, mag	89
		Pt(II)	Pt(2Sura)$_3$$(H_2O)$	ir, uv	91
		Pt(III)	IPt(2Sura)$_4$PtI	X-ray	92
		Pt(IV)	Pt(2Sura)$_2$$Cl_2$	ir, uv	91
		Pd(II)	Pd(2Sura)$_2$	ir, uv	91
		Rh(III)	Rh(2Sura)$_2$Cl	ir, uv	91
		Ti(III)	(MeCp)$_4$$Ti_2$(2Sura)		93
		Tl(III)	$(CH_3)_2$Tl(2Sura)	X-ray	94
		W(0)	$W(CO)_5$(2Sura)		18
		Zn(II)	Zn(gly)(2Sura)(H_2O)	ir	6

TABLE 1 (CONTINUED)

Base no.	Base	Metal	Stoichiometry	Method	Ref.
			***S,S*-2-Thiouracil**		
U6	2SSura	Cu(I)	$Cu_2(2SSura)(OH)_2$		90
			$Cu_2Cl_2(2SSura)$		90
			$Na_2[Cu_2(2SSura)(OH)_2]$		90
			4-Thiouracil		
U7	4Sura	Ag(I)	Ag(4Sura)	ir	95
		Au(III)	$Au_2(4Sura)_2$	ir	95
		Cd(II)	$Cd(4Sura)_2$	ir	95
		Hg(II)	$Hg(4Sura)_2$	ir	95
		Pb(II)	Pb(4Sura)	ir	95
		Tl(I)	Tl(4Sura)	ir	95
			5-Bromouracil		
U8	5Brura	Ag(I)	Ag(5Brura)	pot	1
		Cu(II)	$[Cu(5Brura)]^+$	pot	1
				ion	96
		Ni(II)	$[Ni(5Brura)]^+$	pot	1
		Pt(II)	*cis*-$[(NH_3)_2Pt(DMSO)(5Brura)]^+$	nmr	97
			cis-$(NH_3)_2Pt(5Brura)_2$	nmr	97
			5-Chlorouracil		
U9	5Clura	Ag(I)	Ag(5Clura)	pot	1
		Cu(II)	$[Cu(5Clura)]^+$	pot	1
		Ni(II)	$[Ni(5Clura)]^+$	pot	1
			5-Fluorouracil		
U10	5Fura	Ag(I)	Ag(5Fura)	pot	1
			$K[Ag(5Fura)_2]$	pot	98
			$K_2[Ag_2(5Fura)_3]$		98
			$Ag_2(5Fura)$		98
		Cd(II)	$Cd(OH)_2(5Fura)$	ir, uv	99
		Co(II)	$Co(OH)_2(5Fura)$	ir, uv, mag	99
		Cu(II)	$Cu(OH)_2(5Fura)$	ir, uv, mag	99
			$[Cu(5Fura)]^+$	pot	1, 100
				ion	96
			$Cu(5Fura)_2(H_2O)_2$	ir, mag, spec	100
			Cu(glygly)(5Fura)	ir, esr, mag, uv	101
		Mn(II)	$Mn(OH)_2(5Fura)$	ir, uv, mag	99
		Ni(II)	$Ni(OH)_2(5Fura)_2$	ir, uv, mag	99
			$[Ni(5Fura)]^+$	pot	1, 100

TABLE 1 (CONTINUED)

Base no.	Base	Metal	Stoichiometry	Method	Ref.
			$Ni(5Fura)_2(H_2O)_4$	ir, mag, spec	100
		Pd(II)	$K_2[PdCl_2(5Fura)_2]$		102
				pot	103
			trans-$Pd(NH_3)_2(5Fura)_2$		102
			Pd(en)Cl(5Fura)		102
			$Pd_2(NH_3)_4Cl_2(5Fura)$		102
		Pt(II)	*cis*-$K_2[PtCl_2(5Fura)_2]$		104
			cis-$Pt(NH_3)_2(5Fura)_2$	pot, pK	104
				nmr	97
			trans-$Pt(NH_3)_2(5Fura)_2$		104
			$[Pt(NH_3)_3(5Fura)]NO_3$	pK	104
			cis-$[Pt(NH_3)_2(DMSO)(5Fura)]^+$	nmr	97
		Sn(IV)	$Sn(phen)(Et)_2Cl(5Fura)$	ir, uv, nmr	105
		Zn(II)	$Zn(OH)_2(5Fura)$	ir, uv	99
			$[Zn(5Fura)]^+$	pot	100
			5-Iodouracil		
U11	5Iura	Cu(II)	$[Cu(5Iura)]^+$	ion	96
			5-Nitrouracil		
U12	$5NO_2ura$	Pt(II)	*cis*-$[(NH_3)_2Pt(DMSO)(5NO_2ura)]^+$	nmr	97
			5-Ruracil		
U13	5Rura	Cu(II)	$Cu(5Rura)_2$	X-ray, uv	106
	R = CH=NR′		$Cu(5Rura)_2.2H_3BO_3.2H_2O$	X-ray	107
	R = CH_2-NHR′		0.5EtOH		
			6-Methyluracil		
U14	6meura	Cu(II)	Cu(glygly)(6meura).HBr	ir	14
		Pt(n) (n > 2)	Pt-(6meura)-blue	esr	108
			6-Azauracil		
U15	6azura	Cu(II)	$Cu(6azura)_2(H_2O).2H_2O$	X-ray	109
			Orotic Acid		
U16	oro	Cu(II)	Cu(oro)	pot	61
			cis-$Cu(NH_3)_2(oro)$	X-ray	110
		Co(III)	$[Co(oro)(H_2O)(OH)(NH_3)]_n$	X-ray	111
		Hg(II)	Hg(oro)	pot	61
			$[Hg_2(oro)_3]^{2-}$	pot	61
		Ni(II)	$[Ni(oro)(H_2O)_2(NH_3)]_n$	X-ray	111

TABLE 1 (CONTINUED)

Base no.	Base	Metal	Stoichiometry	Method	Ref.
			$Ni(oro)(H_2O)_2(NH_3)_2$	X-ray	112
			$Ni(oro)(H_2O)_2.H_2O$	X-ray	113, 114
		Pt(II)	*cis*-$(NH_3)_2Pt(oro)$	X-ray	115
				pot	61
			$K_2[Pt(oro)_2]$	uv	116
			cis-$[Pt(oro)_2(NH_3)]^{2-}$	pot	61
		Zn(II)	$Zn(NH_3)_3(oro)$	X-ray	117
			$Zn(oro)(H_2O)_4.H_2O$	X-ray	114
			1,3-Dimethyluracil		
U17	13dme-ura	Cu(II)	$CuCl_2(13dmeura)_2$	X-ray	118
				ir, spec	119
		Cd(II)	$CdCl_2(13dmeura)$	ir, spec	119
			$CdBr_2(13dmeura)$	ir, spec	119
			$CdI_2(13dmeura)_2$	ir, spec	119
		Co(II)	$CoCl_2(13dmeura)_2$	ir, spec	119
			$CoBr_2(13dmeura)_2$	ir, spec	119
		Fe(II)	$FeCl_2(13dmeura)_2$	ir, spec	119
			$FeBr_2(13dmeura)_2$	ir, spec	119
		Mn(II)	$MnCl_2(13dmeura)_2$	ir, spec	119
			$MnBr_2(13dmeura)_2$	ir, spec	119
			$MnI_2(13dmeura)_4$	ir, spec	119
		Ni(II)	$NiCl_2(13dmeura)_2$	ir, spec	119
			$NiBr_2(13dmeura)_2$	ir, spec	119
			$NiI_2(13dmeura)_2$	ir, spec	119
		Pt(III)	$[(13dmeura)Pt(NH_3)_2(1meura)_2Pt(NH_3)_2](NO_3)_3$	nmr	120
			5-Chloro-1-methyluracil		
U18	5C1-1meura	Pt(II)	*cis*-$(NH_3)_2Pt(5Cl1meura)_2$		121
			cis-$[(NH_3)_2Pt(5Cl1meura)_2Pt(NH_3)_2](NO_3)_2$		121
			cis-$(NH_3)_2Pt(5Cl1meura)_2PtCl_2$		121
		Pt(IV)	*cis,mer*-$Pt(NH_3)_2Cl_3(5Cl1meura)$	X-ray	49
		Pt(II)/Cu(II)	$[(NH_3)_2Pt(5Cl1meura)_2Cu(5Cl1meura)_2Pt(NH_3)_2](NO_3)_2$		121

TABLE 1 (CONTINUED)

Base no.	Base	Metal	Stoichiometry	Method	Ref.
			1-Methyl-5-nitrouracil		
U19	1me5NO_2-ura	Pt(II)	*cis*-$(NH_3)_2$Pt(1me5NO_2ura)$_2$		121
			2,4-Dithiouracil		
U20	24dSura	Ag(I)	Ag_2(24dSura)	ir	122
		Au(III)	Au_2(24dSura)$_3$	ir	122
		Cd(II)	Cd(24dSura)	ir	122
		Co(II)	Co_4(24dSura)$_9$$(H_2O)_2$	ir, uv, mag	122
		Cu(II)	Cu(24dSura)$_2$	ir, uv, mag	122
		Ir(III)	$IrCl_3$(24dSura)$_3$	ir, xps	123
		Ni(II)	Ni(24dSura)$_2$$(H_2O)_4$	ir, mag	122
			Ni(24dSura)(py)$_2$	ir, uv, mag	122
		Pb(II)	Pb(24dSura)	ir	122
		Pd(0)	Pd$(PPh_3)_2$(24dSura)$_2$	ir	124
		Pd(II)	Pd(24dSura)$_2$	ir, xps	123
				uv	125
		Pd(IV)	Pd(24dSura)$_5$$Cl_2$	ir, uv	125
		Pt(II)	Pt(24dSura)$_2$	ir, xps	123
				uv, ir	125
		Pt(IV)	Pt(24dSura)$_2$$Cl_2$	uv, ir	125
		Rh(I)	$Rh_3(PPh_3)_2Cl_2$(24dSura)$_4$	ir	124
		Rh(II)	Rh_2(24dSura)$_4$$(H_2O)_2$	ir	124
		Rh(III)	$RhCl_3$(24dSura)$_3$	ir, xps	123
			RhCl(24dSura)$_2$	ir, uv	125
		Ru(II)	$Ru_2(H_2O)_2$(24dSura)$_3$	ir, uv	124
		Ru(III)	Ru_2(24dSura)$_5$Cl	uv, ir, mag	125
		Ti(III)	$(MeCp)_2Ti_2$(24dSura)	X-ray	93
		Tl(I)	Tl(24dSura)$(H_2O)_2$	ir	122
			6-Methyl-2-thiouracil		
U21	6me2-Sura	Cu(I)	Cu(6me2Sura).HCl		90
		Ir(III)	[Ir(6me2Sura)$_3$]Cl_3	ir, xps	126
		Pd(II)	Pd(6me2Sura)$_2$	ir, xps	126
		Pt(II)	Pt(6me2Sura)$_2$	ir, xps	126
		Rh(III)	[Rh(6me2Sura)$_3$]Cl_3	ir, xps	126
			6-Amino-2-thiouracil		
U22	6A2-Sura	Cu(I)	Cu(6A2Sura).2HCl		90
			3-Methylorotic acid		
U23	3meoro	Pt(II)	K_2[Pt(3meoro)$_2$]	uv	116

TABLE 1 (CONTINUED)

Base no.	Base	Metal	Stoichiometry	Method	Ref.
			5,6-Dihydrouracil		
U24	56dHura	Hg(II)	$HgCl_2(56dHura)_2$	X-ray	15
			5,6-Diaminouracil		
U25	56dAura	Cu(II)	$[Cu(56dAura)_2]SO_4$		127
		Ni(II)	$[Ni(56dAura)_2(H_2O)_2]SO_4$		127
			5-Nitroorotic acid		
U26	$5NO_2oro$	Cu(II)	$Cu(NH_3)_2(5NO_2oro).H_2O$	X-ray, uv	128
			$Cu_3(NH_3)_6(5NO_2oro)_2.5H_2O$	X-ray, uv	128
		Pt(II)	$K_2[Pt(5NO_2oro)_4]$	uv	116
			$(NH_3)_2Pt(5NO_2oro)$	uv	116
			$K_2[Pt(5NO_2oro)_2]$	uv	116
			5,6-Dihydro-5,6-dihydroxythymine		
U27	56dH-dHOthy	Os(VI)	$Os(py)_2O_2(56dHdHOthy)$	X-ray	129
			6-Amino-5-nitrouracil		
U28	$6A5NO_2$-ura	Cd(II)	$Cd(6A5NO_2ura)_2$	therm	130
		Co(II)	$Co(6A5NO_2ura)_2$	therm	131
		Cu(II)	$Cu(6A5NO_2ura)_2$	therm	131
		Fe(II)	$Fe(6A5NO_2ura)_2$	therm	131
		Hg(II)	$Hg_3Cl_6(6A5NO_2ura)_4$	therm	132
		Ni(II)	$Ni(6A5NO_2ura)_2$	therm	131
		Zn(II)	$Zn(6A5NO_2ura)_2(H_2O)_2$	therm	133
			6-Amino-5-formyluracil		
U29	6A5fura	Cu(II)	$Cu(6A5fura)_2$	therm	134
		Ni(II)	$Ni(NH_3)_2(6A5fura)_2$	therm	134
		Pd(II)	$Pd(6A5fura)_6$	therm	134
			Dithymine		
U30	dthy	Cu(II)	$[Cu(dthy)]^+$	pot	61
			Cu(dthy)	pot	61
		Hg(II)	Hg(dthy)	pot	61
			$[Hg(dthy)]^-$	pot	61
			$[Hg(dthy)]^{2-}$	pot	61

TABLE 1 (CONTINUED)

Base no.	Base	Metal	Stoichiometry	Method	Ref.
			Diorotic Acid		
U31	doro	Cu(II)	$[Cu(doro)]^{2-}$	pot	61
		Hg(II)	$[Hg(doro)]^{2-}$	pot	61
		Pt(II)	*cis*-$[Pt(NH_3)_2(doro)]^{2-}$	pot	61
			5,6-Dihydro-1-methyluracil		
U32	56dH-1meura	Pt(II)	*cis*-$[Pt(NH_3)_2(56dH1meura)_2]NO_3$		121
			5,5-Dichloro-6-hydroxy-5,6-dihydro-1-methyluracil		
U33	55dCl6HO-56dH1meura	Pt(IV)	*cis,mer*-$Pt(NH_3)_2Cl_3$(55dCl6HO 56dH1meura)	X-ray	49
			1,3-Dialkyl-5-fluorouracil		
U34	13dR5-Fura (R = me, et, pr)	Pd(II)	*cis*-$(PPh_3)_2$Pd(13dR5Fura)I	ir, nmr	135
			trans-$(PPh_3)_2$Pd(13dR5Fura)I	ir, nmr	135
		Pd(II)	Pd(bpy)(13dR5Fura)I	ir, nmr	135
			Pd(diphos)(13dR5Fura)I	ir, nmr	135
			6-Amino-1-methyl-5-nitrosouracil		
U35	6A1me-5NOura	Cd(II)	$Cd(6A1me5NOura)_2Cl_2$	therm	130
		Hg(II)	$Hg(6A1me5NOura)_2$	therm	132
		Zn(II)	$Zn(6A1me5NOura)_2$	therm	133
			$Zn(6A1me5NOura)_2(H_2O)_2$	therm	133
			6-Amino-3-methyl-5-nitrosouracil		
U36	6A3me 5NOura	Cd(II)	$Cd(6A3me5NOura)_2$	X-ray	136
				ir	137
				therm	130
		Hg(II)	$HgCl_2(6A3me5NOura)_2$	therm	132
				ir	137
		Zn(II)	$Zn(6A3me5NOura)_2$	therm	133
				ir	137
			6-Amino-1,3-dimethyl-5-nitrosouracil		
U37	6A13dme-5NOura	Co(II)	$Co(6A13dme5NOura)_3$	ir	138
		Cu(II)	$Cu(6A13dme5NOura)_2$	ir	138
		Ni(II)	$Ni(6A13dme5NOura)_2$	ir	138
		Pd(II)	$Pd(6A13dme5NOura)_2$	ir	138
		Pt(II)	$Pt(6A13dme5NOura)_2$	ir	138

TABLE 1 (CONTINUED)

Base no.	Base	Metal	Stoichiometry	Method	Ref.
			6-Amino-5-formyl-1-methyluracil		
U38	6A5f-1meura	Cu(II)	$Cu(6A5f1meura)_2$	therm	134
		Ni(II)	$Ni(NH_3)_2(6A5f1meura)_2$	therm	134
		Pd(II)	$Pd(6A5f1meura)_2$	therm	134
			6-Amino-5-formyl-3-methyluracil		
U39	6A5f-3meura	Cu(II)	$Cu(6A5f3meura)_2$	therm	138
		Ni(II)	$Ni(NH_3)_2(6A5f3meura)_2$	therm	138
		Pd(II)	$Pd(6A5f3meura)_2$	therm	138
			6-Amino-1,3-dimethyl-5-formyluracil		
U40	6A13dme-5fura	Cu(II)	$Cu(6A13dme5fura)_2$	therm	138
		Ni(II)	$[Ni(6A13dme5fura)_2]NO_3$	therm	138
		Pd(II)	$Pd(6A13dme5fura)_2$	therm	138
			5,6-Dihydro-5,6-dihydroxy-1-methylthymine		
U41	56dHdHO=1methy	Os(VI)	$Os(py)_2O_2(56dHdHO1methy).H_2O.0.5py$	X-ray	139

TEXT REFERENCES

1. **G. L. Eichhorn and L. G. Marzilli, Eds.,** *Metal Ions in Genetic Information Transfer,* Elsevier/North-Holland, New York, 1981.
2. **B. Rosenberg, L. Van Camp, J. E. Trosko, and V. H. Mansour,** *Nature (London),* 1969, *222,* 385.
3. **M. Nicolini, Ed.,** *Platinum and Other Metal Coordination Compounds in Cancer Chemotherapy,* Martinus Nijhoff, Boston, 1988.
4. **L. G. Marzilli,** *Prog. Inorg. Chem.,* 1977, *23,* 255.
5. **T. G. Spiro, Ed.,** *Nucleic Acid-Metal Ion Interactions,* John Wiley & Sons, New York, 1980.
6. **S. J. Lippard,** *Acc. Chem. Res.,* 1978, *11,* 211.

7a. **R. B. Martin and Y. H. Mariam,** *Met. Ions Biol. Syst.,* 1979, *8,* 57.

7b. **R. B. Martin,** *Acc. Chem. Res.,* 1985, *18,* 32.

8. **V. Swaminathan and M. Sundaralingam,** *CRC Crit. Rev. Biochem.,* 1979, *6,* 245.
9. **R. W. Gellert and R. Bau,** *Met. Ions Biol. Syst.,* 1979, *8,* 1.
10. **B. de Castro, T. J. Kistenmacher, and L. G. Marzilli,** in *Trace Elements in the Pathogenesis and Treatment of Inflammatory Conditions,* Vol. 8, K. D. Raindsford, K. Brune, and M. W. Whitehouse, Eds., Birkhäuser, Basel, 1981, 434.
11. **J. Reedijk, A. M. J. Fichtinger-Schepman, A. T. van Oosterom, and P. van de Putte,** *Struct. Bond.,* 1987, *67,* 53.
12. **D. J. Brown,** in *Comprehensive Heterocyclic Chemistry,* Vol. 3, A. R. Katritzky and C. Rees, Eds., Pergamon Press, Oxford, 1983, 57.

13a. **R. Wagner and W. von Philipsborn,** *Helv. Chim. Acta,* 1970, *53,* 299.

13b. **A. R. Katritzky and A. J. Waring,** *J. Chem. Soc.,* 1962, 1540.

14a. **W. N. Hunter, T. Brown, N. N. Anand, and O. Kennard,** *Nature (London),* 1986, *320,* 552.

14b. **G. J. Quigley, G. Ughetto, G. A. van der Marel, J. H. van Boom, A. H.-J. Wang, and A. Rich,** *Science,* 1986, *232,* 1255.

15. **O. N. Voloshin, S. M. Mirkin, V. I. Lyamichev, B. P. Belotserkovskii, and M. D. Frank-Kamenetskii,** *Nature (London),* 1988, *333,* 475.
16. **R. Taylor and O. Kennard,** *J. Mol. Struct.,* 1982, *78,* 1.
17. **J. Bandekar and G. Zundel,** *Spectrochim. Acta Part A,* 1983, *39A,* 343.

18a. **J. S. Kwiatkowski and B. Pullman,** *Adv. Heterocycl. Chem.,* 1975, *18,* 199.

18b. **M. J. Scanlan and I. H. Hillier,** *J. Am. Chem. Soc.,* 1984, *106,* 3737.

18c. **P. Beak and J. M. White,** *J. Am. Chem. Soc.,* 1982, *104,* 7073.

19a. **J. J. Christensen, J. H. Rytting, and R. M. Izatt,** *J. Phys. Chem.,* 1967, *71,* 2700.

19b. **D. Shugar and J. J. Fox,** *Biochim. Biophys. Acta,* 1952, *9,* 199.

19c. **J. R. De Member and F. A. Wallace,** *J. Am. Chem. Soc.,* 1975, *97,* 6240.

20. **K. Nakanishi, N. Suzuki, and F. Yamazaki,** *Bull. Chem. Soc. Jpn.,* 1961, *34,* 53.
21. **K. L. Wierzchowski, E. Litonska, and D. Shugar,** *J. Am. Chem. Soc.,* 1965, *87,* 4621.
22. **B. Lippert,** *J. Raman Spectrosc.,* 1979, *8,* 274.
23. **C. J. L. Lock, P. Pilon, and B. Lippert,** *Acta Crystallogr. Sect. B,* 1979, *B35,* 2533.
24. **L. C. Snyder, R. G. Shulman, and D. B. Neumann,** *J. Chem. Phys.,* 1970, *53,* 256.
25. **P. Lumme and I. Mutikainen,** *Acta Crystallogr. Sect. B,* 1980, *B36,* 2251.
26. **I. Mutikainen and P. Lumme,** *Acta Crystallogr. Sect. B,* 1980, *B36,* 2237.
27. **I. Mutikainen,** Ph. D. Thesis, University of Helsinki, Finland, 1988.
28. **K. Inagaki and Y. Kidani,** *Bioinorg. Chem.,* 1978, *9,* 157.
29. **B. Lippert,** *Inorg. Chem.,* 1981, *20,* 4326.
30. **R. Pfab, P. Jandik, and B. Lippert,** *Inorg. Chim. Acta,* 1982, *66,* 193.
31. **C. J. L. Lock and B. Lippert,** unpublished results.
32. **B. F. Fieselmann, D. N. Hendrickson, and G. D. Stucky,** *Inorg. Chem.,* 1978, *17,* 1841.

33a. **B. Lippert and D. Neugebauer,** *Inorg. Chim. Acta,* 1980, *46,* 171.
33b. **F. Guay, A. L. Beauchamp, C. Gilbert, and R. Savoie,** *Can. J. Spectrosc.,* 1983, *28,* 13.
34a. **H. Susi, J. S. Ard, and J. M. Purcell,** *Spectrochim. Acta Part A,* 1973, *29A,* 725.
34b. **H. Morita and S. Nagakura,** *Theor. Chim. Acta,* 1986, *11,* 279.
35a. **M. J. Scanlan and I. H. Hillier,** *J. Chem. Soc. Chem. Commun.,* 1984, 102.
35b. **M. Dreyfus, O. Bensaude, G. Dodin, and J. E. Dubois,** *J. Am. Chem. Soc.,* 1976, *98,* 6338.
36. **M. G. Harris and R. Stewart,** *Can. J. Chem.,* 1977, *55,* 3800.
37. **B. Lippert,** *Inorg. Chim. Acta,* 1981, *55,* 5.
38. **B. Lippert, H. Schöllhorn, and U. Thewalt,** *J. Am. Chem. Soc.,* 1989, *111,* 7213.
39. **B. Lippert, H. Schöllhorn, and U. Thewalt,** *J. Am. Chem. Soc.,* 1986, *108,* 6616.
40. **R. Faggiani, B. Lippert, C. J. L. Lock, and R. A. Speranzini,** *J. Am. Chem. Soc.,* 1981, *103,* 1111.
41. **B. J. Graves and D. J. Hodgson,** *J. Am. Chem. Soc.,* 1979, *101,* 5608.
42. **J.-P. Charland, M. Simard, and A. L. Beauchamp,** *Inorg. Chim. Acta,* 1983, *80,* L57.
43. **R. M. K. Dale, D. C. Livingston, and D. C. Ward,** *Proc. Natl. Acad. Sci. U.S.A.,* 1973, *70,* 2238.
44. **H. Schöllhorn, U. Thewalt, and B. Lippert,** *J. Chem. Soc. Chem. Commun.,* 1986, 258.
45. **B. Lippert, U. Thewalt, H. Schöllhorn, D. M. L. Goodgame, and R. W. Rollins,** *Inorg. Chem.,* 1984, *23,* 2807.
46. **E. Buncel, C. Boone, and H. Holy,** *Inorg. Chim. Acta,* 1986, *125,* 167.
47. **W. Micklitz, J. Riede, B. Huber, G. Muller, and B. Lippert,** *Inorg. Chem.,* 1988, *27,* 1979 (and references cited therein).
48. **F. Guay and A. Beauchamp,** *Inorg. Chim. Acta,* 1982, *66,* 57.
49. **H. Schöllhorn, U. Thewalt, and B. Lippert,** *Inorg. Chim. Acta,* 1987, *135,* 155.
50. **M. Goodgame and D. A. Jakubovic,** *Coord. Chem. Rev.,* 1987, *79,* 97.
51. **B. Lippert,** in *Metal-Based Anti-Tumour Drugs,* M. Gielen, Ed., Freund, London, 1988, 201.
52. **B. L. Kindberg, E. H. Griffith, and E. L. Amma,** *J. Chem. Soc. Chem. Commun.,* 1975, 195.
53. **J. J. Fiol, A. Terron, D. Mulet, and V. Moreno,** *Inorg. Chim. Acta,* 1987, *135,* 197.

TABLE REFERENCES

1. **Y. L. Tan and A. Beck,** *Biochem. Biophys. Acta,* 1973, *299,* 500.
2. **J. R. De Member and F. A. Wallace,** *J. Am. Chem. Soc.,* 1975, *97,* 6240.
3. **M. Goodgame and K. W. Johns,** *J. Chem. Soc. Dalton Trans.,* 1977, *17,* 1680.
4. **I. Mutikainen and P. Lumme,** *Acta Crystallogr. Sect. B,* 1980, *B36,* 2237.
5. **A. R. Sarkar and P. Gosh,** *Inorg. Chim. Acta,* 1983, *78,* L39.
6. **M. Gupta and M. N. Srivastava,** *Polyhedron,* 1984, *4,* 475.
7. **M. M. Taqui Khan and C. R. Krishnamoorthy,** *J. Inorg. Nucl. Chem.,* 1974, *36,* 711.
8. **P. Gosh, T. K. Mukhopadhyay, and A. R. Sarkar,** *Trans. Met. Chem.,* 1984, *9,* 46.
9. **M. M. Taqui Khan and S. Satyanarayana,** *Indian J. Chem.,* 1981, *20,* 814.
10. **M. M. Taqui Khan, S. Satyanarayana, M. S. Jyoti, and A. P. Reddy,** *Indian J. Chem. Sect. A,* 1983, *22A,* 364.
11. **N. B. Nigam, P. C. Sinha, and M. N. Srivastava,** *Indian J. Chem. Sect. A,* 1983, *22A,* 818.
12. **N. B. Nigam, P. C. Sinha, M. Gupta, and M. N. Srivastava,** *Indian J. Chem. Sect. A,* 1985, *24A,* 893.
13. **V. Kothekar and S. Dutta,** *Int. J. Quantum Chem.,* 1977, *12,* 505.
14. **T. Fujita, H. Masuno, and T. Sakaguchi,** *Chem. Pharm. Bull.,* 1978, *26,* 2826.
15. **J. A. Carrabine and M. Sundaralingam,** *Biochemistry,* 1971, *10,* 292.
16. **S. Mansy and R. S. Tobias,** *Inorg. Chem.,* 1975, *14,* 287.

17. **R. C. Srivastava and M. N. Srivastava,** *J. Inorg. Nucl. Chem.*, 1978, *40,* 1439.
18. **W. Beck and N. Kottmair,** *Chem. Ber.*, 1976, *109*, 970.
19. **A. M. Bhandari, A. K. Solanki, and S. Wadhwa,** *J. Inorg. Nucl. Chem.*, 1981, *43,* 2995.
20. **J. J. Fiol, A. Terron, and V. Moreno,** *Inorg. Chim. Acta,* 1986, *125,* 159.
21. **R. Weiss and H. Venner,** *Hoppe-Seyler's Z. Physiol. Chem.*, 1969, *350,* 396.
22. **P. Lumme and I. Mutikainen,** *Acta Crystallogr. Sect. B,* 1980, *B36*, 2251.
23. **M.-C. Lim,** *J. Inorg. Nucl. Chem.*, 1981, *43,* 221.
24. **R. Faggiani, B. Lippert, and C. J. L. Lock,** *Inorg. Chem.*, 1980, *19,* 295.
25. **K. Inagaki and Y. Kidani,** *Bioinorg. Chem.*, 1978, *9,* 157.
26. **B. Lippert,** *Inorg. Chem.*, 1981, *20,* 4326.
27. **B. Lippert and C. J. L. Lock,** unpublished results.
28. **B. Lippert,** *J. Clin. Hematol. Oncol.*, 1977, *7,* 26.
29. **H. Neubacher, M. Seul, and W. Lohmann,** *Z. Naturforsch. Teil C,* 1982, *37C*, 553.
30. **J. P. Davidson, P. J. Faber, R. G. Fischer, Jr., S. Mansy, H. J. Peresie, B. Rosenberg, and L. van Camp,** *Cancer Chemother. Rep. Part 1,* 1975, *59,* 287.
31. **W. Bauer, S. L. Gonias, S. K. Kam, K. C. Wu, and S. J. Lippard,** *Biochemistry,* 1978, *17,* 1060.
32. **B. F. Fieselmann, D. N. Hendrickson, and G. D. Stucky,** *Inorg. Chem.*, 1978, *17,* 1841.
33. **K. Aoki and W. Saenger,** *Acta Crystallogr. Sect. C,* 1984, *C40*, 775.
34. **B. Lippert, D. Neugebauer, and G. Raudaschl,** *Inorg. Chim. Acta,* 1983, *78,* 161.
35. **H. Schollhorn, P. Eisenmann, U. Thewalt, and B. Lippert,** *Inorg. Chem.*, 1986, *25,* 3384.
36. **D. Neugebauer and B. Lippert,** *J. Am. Chem. Soc.*, 1982, *104*, 6596.
37. **J. D. Wollins and B. Rosenberg,** *J. Inorg. Biochem.*, 1983, *19,* 41.
38. **B. Lippert, U. Thewalt, H. Schöllhorn, D. M. L. Goodgame, and R. W. Rollins,** *Inorg. Chem.*, 1984, *23,* 2807.
39. **W. Micklitz, J. Riede, B. Huber, G. Müller, and B. Lippert,** *Inorg. Chem.*, 1988, *27,* 1979.
40. **R. Faggiani, C. J. L. Lock, R. J. Pollock, B. Rosenberg, and G. Turner,** *Inorg. Chem.*, 1981, *20,* 804.
41. **G. Raudaschl-Sieber and B. Lippert,** *Inorg. Chem.*, 1985, *24,* 2426.
42. **T. Ramstad, J. D. Wollins, and M. J. Weaver,** *Inorg. Chim. Acta,* 1986, *124*, 187.

43a. **T. V. O'Halloran, P. K. Mascharak, I. D. Williams, M. M. Roberts, and S. J. Lippard,** *Inorg. Chem.*, 1987, *26,* 1261.

43b. **P. K. Mascharak, I. D. Williams, and S. J. Lippard,** *J. Am. Chem. Soc.*, 1984, *106*, 6428.

44. **B. Lippert and D. Neugebauer,** *Inorg. Chem.*, 1982, *21,* 451.
45. **B. Lippert, H. Schöllhorn, and U. Thewalt,** *Z. Naturforsch. Teil B,* 1983, *38B,* 1441.
46. **B. Lippert, H. Schöllhorn, and U. Thewalt,** *J. Am. Chem. Soc.*, 1986, *108,* 525.
47. **B. Lippert, H. Schöllhorn, and U. Thewalt,** *Inorg. Chem.*, 1986, *25,* 407.
48. **H. Schöllhorn, U. Thewalt, and B. Lippert,** *J. Chem. Soc. Chem. Commun.*, 1986, 258.
49. **G. Müller, J. Riede, R. Beyerle-Pfnür, and B. Lippert,** *J. Am. Chem. Soc.*, 1984, *106,* 7999.
50. **H. Schöllhorn, U. Thewalt, and B. Lippert,** *J. Chem. Soc. Chem. Commun.*, 1984, 769.
51. **D. M. L. Goodgame, R. W. Rollins, and B. Lippert,** *Polyhedron,* 1985, *4,* 829.
52. **U. Thewalt, D. Neugebauer, and B. Lippert,** *Inorg. Chem.*, 1984, *23,* 1713.
53. **B. Lippert, H. Schöllhorn, and U. Thewalt,** *Inorg. Chem.*, 1987, *26,* 1736.
54. **H. Schöllhorn, U. Thewalt, and B. Lippert,** *Inorg. Chim. Acta,* 1987, *135,* 155.
55. **D. M. L. Goodgame, M. A. Hitchman, and B. Lippert,** *Inorg. Chem.*, 1986, *25,* 2191.
56. **D. M. Duckworth, D. M. L. Goodgame, M. A. Hitchman, B. Lippert, and K. S. Murray,** *Inorg. Chem.*, 1987, *26,* 1823.
57. **I. Mutikainen, O. Orama, A. Pajunen, and B. Lippert,** *Inorg. Chim. Acta,* 1987, *137,* 189.

58a. **W. Micklitz, G. Müller, J. Riede, and B. Lippert,** *J. Chem. Soc. Chem. Commun.*, 1987, 76.

58b. **W. Micklitz, G. Müller, B. Huber, J. Riede, F. Rashwan, J. Heinze, and B. Lippert,** *J. Am. Chem. Soc.,* 1988, *110,* 7084.
59. **H. Schöllhorn, U. Thewalt, and B. Lippert,** *Inorg. Chim. Acta,* 1985, *108,* 77.
60. **M. M. Taqui-Khan and S. Satyanarayana,** *Indian J. Chem. Sect. A,* 1982, *21A,* 913.
61. **R. M'Boungou, M. Petit-Ramel, G. Thomas-David, G. Perichet, and B. Pouget,** *Can. J. Chem.,* 1987, *65,* 1479.
62. **T. J. Kistenmacher and L. G. Marzilli,** *Inorg. Chem.,* 1975, *14,* 2479.
63. **H. Basch, M. Krauss, W. J. Stevens, and D. Cohen,** *Inorg. Chem.,* 1986, *25,* 684.
64. **R. Pfab, P. Jandik, and B. Lippert,** *Inorg. Chim. Acta,* 1982, *66,* 193.
65. **J. D. Woollins and B. Rosenberg,** *Inorg. Chem.,* 1982, *21,* 1280.
66. **B. Lippert, R. Pfab, and D. Neugebauer,** *Inorg. Chim. Acta,* 1979, *37,* L495.
67. **B. Lippert,** *J. Raman Spectrosc.,* 1980, *9,* 324.
68. **R. Faggiani, B. Lippert, C. J. L. Lock, and R. Pfab,** *Inorg. Chem.,* 1981, *20,* 2381.
69. **P. Zaplatynski, H. Neubacher, and W. Lohmann,** *Z. Naturforsch. Teil B,* 1979, *34b,* 1466.
70. **F. Guay and A. L. Beauchamp,** *J. Am. Chem. Soc.,* 1979, *101,* 6260.
71. **F. Guay, A. L. Beauchamp, C. Gilbert, and R. Savoie,** *Can. J. Spectrosc.,* 1983, *28,* 13.
72. **R. Faggiani, H. E. Howard-Lock, C. J. L. Lock, and M. A. Turner,** *Can. J. Chem.,* 1987, *65,* 1568.
73. **L. D. Kosturko, C. Folzer, and R. F. Stewart,** *Biochemistry,* 1974, *13,* 3949.
74. **R. W. Chrisman, S. Mansy, H. J. Peresie, A. Ranade, T. A. Berg, and R. S. Tobias,** *Bioinorg. Chem.,* 1977, *7,* 245.
75. **F. Guay and A. L. Beauchamp,** *Inorg. Chim. Acta,* 1982, *66,* 57.
76. **H. Schöllhorn, U. Thewalt, and B. Lippert,** *Inorg. Chim. Acta,* 1985, *106,* 177.
77. **B. Lippert and D. Neugebauer,** *Inorg. Chim. Acta,* 1980, *46,* 171.
78. **B. Lippert,** *Inorg. Chim. Acta,* 1981, *55,* 5.
79. **R. Beyerle and B. Lippert,** *Inorg. Chim. Acta,* 1982, *66,* 141.
80. **C. J. L. Lock, H. J. Peresie, B. Rosenberg, and G. Turner,** *J. Am. Chem. Soc.,* 1978, *100,* 3371.
81. **D. Neugebauer and B. Lippert,** *Inorg. Chim. Acta,* 1982, *67,* 151.
82a. **H. Schöllhorn, U. Thewalt, and B. Lippert,** *Inorg. Chim. Acta,* 1984, *93,* 19.
82b. **B. Lippert, D. Neugebauer, and U. Schubert,** *Inorg. Chim. Acta,* 1980, *46,* L11.
83. **W. Micklitz, O. Renn, H. Schöllhorn, U. Thewalt, and B. Lippert,** *Inorg. Chem.,* 1990, *29,* in press.
84. **A. J. Thomson, I. A. G. Roos, and R. D. Graham,** *J. Clin. Hematol. Oncol.,* 1977, *7,* 242.
85. **B. Lippert and D. Neugebauer,** *Inorg. Chim. Acta,* 1980, *46,* 171.
86. **B. Lippert and U. Schubert,** *Inorg. Chim. Acta,* 1981, *56,* 15.
87. **D. M. L. Goodgame, R. W. Rollins, and B. Lippert,** *Inorg. Chim. Acta,* 1982, *66,* L59.
88a. **C. W. Hunt and E. L. Amma,** *J. Chem. Soc. Chem. Commun.,* 1973, 869.
88b. **C. W. Hunt, E. A. H. Griffith, and E. L. Amma,** *Inorg. Chem.,* 1976, *15,* 2993.
89. **I. P. Khullar and U. Agarwala,** *Aust. J. Chem.,* 1974, *27,* 1877.
90. **R. Weiß and H. Venner,** *Z. Anorg. Allg. Chem.,* 1959, *317,* 82.
91. **I. P. Khullar and U. Agarwala,** *Aust. J. Chem.,* 1975, *28,* 1529.
92a. **D. M. L. Goodgame, R. W. Rollins, and A. C. Skapski,** *Inorg. Chim. Acta,* 1984, *83,* L11.
92b. **D. M. L. Goodgame, R. W. Rollins, A. M. Z. Slawin, D. J. Williams, and P. W. Zard,** *Inorg. Chim. Acta,* 1986, *120,* 91.
93. **D. R. Corbin, L. C. Francesconi, D. N. Hendrickson, and G. D. Stucky,** *J. Chem. Soc. Chem. Commun.,* 1979, 248.
94. **M. S. Garcia-Tasende, M. I. Suarez, A. Sanchez, J. S. Casas, J. Sordo, E. E. Castellano, and Y. P. Mascarenhas,** *Inorg. Chem.,* 1987, *26,* 3813.
95. **I. P. Khullar and U. Agarwala,** *Ind. J. Chem.,* 1974, *12,* 1096.
96. **L. Kaminski and H. Altmann,** *Experientia,* 1967, *23,* 599.

97. **D. N. Bochkov, N. A. Smorygo, L. L. Shcherbakova, and B. A. Ivin,** *Zh. Obshch. Khim.*, 1984, *54*, 900.
98. **M. I. Gel'fman and N. A. Kustova,** *Russ. J. Inorg. Chem.*, 1970, *15*, 47.
99. **U. P. Singh, R. Ghose, and A. K. Ghose,** *Inorg. Chim. Acta,* 1987, *136*, 21.
100. **M. Aplincourt, A. Debras, and C. Gerard,** *J. Chem. Res. S,* 1985, 24.
101. **J. F. Villa, J. Gelber, N. Khoe, and J. Cepeda,** *J. Am. Chem. Soc.*, 1978, *100,* 4305.
102. **M. I. Gel'fman and N. A. Kustova,** *Russ. J. Inorg. Chem.*, 1969, *14*, 985.
103. **M. I. Gel'fman and N. A. Kustova,** *Russ. J. Inorg. Chem.*, 1969, *14,* 1113.
104. **M. I. Gel'fman and N. A. Kustova,** *Russ. J. Inorg. Chem.*, 1970, *15,* 1602.
105. **W. Jiazhu, H. Jingshuo, H. Liyiao, S. Dashuang, and H. Sengzhi,** *Inorg. Chim. Acta,* 1988, *152*, 67.
106. **I. Sasaki, M. N. Dufour, A. Gaudemer, A. Chiaroni, C. Riche, D. Parker-Decrouez, and P. Boucly,** *Nouv. J. Chim.*, 1984, *8*, 237.
107. **I. Sasaki, M. N. Dufour, A. Gaudemer, A. Chiaroni, and C. Riche,** *Inorg. Chim. Acta,* 1986, *112*, 129.
108. **M. Seul, H. Neubacher, and W. Lohmann,** *Z. Naturforsch., Teil B,* 1981, *36B*, 651.
109. **B. A. Mosset, J. J. Bonnet, and J. Galy,** *Acta Crystallogr. Sect. B,* 1977, *B33,* 2639.
110. **I. Mutikainen and P. Lumme,** *Acta Crystallogr. Sect. B,* 1980, *B36,* 2233.
111. **I. Mutikainen,** *Finn. Chem. Lett.*, 1985, 193.
112. **P. Lumme and I. Mutikainen,** *Acta Crystallogr. Sect. B,* 1980, *B36*, 2251.
113. **M. Sabat, D. Zglinska, and B. Jezowska-Trzebiatowska,** *Acta Crystallogr. Sect. B,* 1987, *B36,* 1187.
114. **A. Karipides and B. Thomas,** *Acta Crystallogr. Sect. C,* 1986, *C42,* 1705.
115. **T. Solin, K. Matsumoto, and K. Fuwa,** *Bull. Chem. Soc. Jpn.*, 1981, *54,* 3731.
116. **P. Arrizabalaga, P. Castan, and J. P. Laurent,** *Inorg. Chim. Acta,* 1984, *92,* 203.
117. **I. Mutikainen,** *Inorg. Chim. Acta*, 1987, *136,* 155.
118. **B. A. Cartwright, M. Goodgame, K. W. Johns, and A. C. Skapski,** *Biochem. J.,* 1978, *175*, 337.
119. **M. Goodgame and K. W. Johns,** *Inorg. Chim. Acta,* 1978, *30,* L335.
120. **F. Pesch and B. Lippert,** unpublished results.
121. **O. Krizanovic and B. Lippert,** in *Platinum and Other Metal Coordination Compounds in Cancer Chemotherapy,* M. Nicolini, Ed., Martinus Nijhoff, Boston, 1988, 700.
122. **J. S. Dwivedi and U. Agarwala,** *Indian J. Chem.*, 1972, *10*, 652.
123. **J. R. Lusty, H. S. O. Chan, and J. Peeling,** *Trans. Met. Chem.*, 1983, *8,* 343.
124. **J. S. Dwivedi and U. Agarwala,** *Z. Anorg. Allg. Chem.,* 1973, *397*, 74.
125. **J. S. Dwivedi and U. Agarwala,** *Indian J. Chem.*, 1972, *10,* 657.
126. **J. R. Lusty, J. Peeling, and M. A. Abdel-Aal,** *Inorg. Chim. Acta,* 1981, *56,* 21.
127. **R. Weiß and F. Hein**, *Hoppe-Seyler's Z. Physiol. Chem.*, 1959, *317*, 95.
128. **P. Arrizabalaga, P. Castan, and F. Dahan**, *Inorg. Chem.*, 1983, *22*, 2245.
129. **S. Neidle and D. I. Stuart,** *Biochem. Biophys. Acta,* 1976, *418*, 226.
130. **M. N. Moreno-Carretero and J. M. Salas-Peregrin,** *J. Therm. Anal.,* 1984, *29,* 1053.
131. **J. M. Salas-Peregrin, M. A. Romero-Molina, M. A. Ferro-Garcia, and M. A. Romero-Carretero,** *J. Therm. Anal.,* 1985, *30,* 921.
132. **M. N. Moreno-Carretero and J. M. Salas-Peregrin,** *J. Therm. Anal.*, 1985, *30,* 377.
133. **M. N. Moreno-Carretero, J. M. Salas-Peregrin, and A. Mata-Arjona,** *J. Therm. Anal.*, 1984, *29,* 553.
134. **L. E. Garcia-Megias, E. Colacio-Rodriguez, A. Garcia-Rodriguez, J. Ruiz-Sanches, and J. M. Salas-Peregrin,** *J. Therm. Anal.*, 1987, *32,* 1127.
135. **H. Urata, M. Tanaka, and T. Fuchikami,** *Chem. Lett.*, 1987, 751.
136. **M. A. Romero, M. N. Moreno, J. Ruiz, M. P. Sanchez, and F. Nieto,** *Inorg. Chem.,* 1986, *25,* 1498.
137. **J. M. Salas-Peregrin, M. N. Moreno-Carretero, and E. Colacio-Rodriguez,** *Can. J. Chem.,* 1985, *63,* 3573.

138. **G. Cruz-Bermudez, A. Garcia-Rodriguez, M. Moreno-Carretero, J. M. Salas-Peregrin, and C. Valenzuela-Calahorro,** *Monatsh. Chem.,* 1987, *118,* 329.
139. **T. J. Kistenmacher, L. G. Marzilli, and M. Rossi,** *Bioinorg. Chem.,* 1976, *6,* 347.
140. **M. Bressan, R. Ettore, and P. Rigo,** *J. Magn. Reson.,* 1977, *26,* 43.
141. **Y. M. Temerk, M. M. Kamal, M. E. Ahmed, and M. I. Abd-el-Hamid,** *Bioelectrochem. Bioenerg.,* 1984, *12,* 475.
142. **R. Weiss and H. Venner,** *Hoppe Seyler's Z. Physiol. Chem.,* 1965, *341,* 229.
143. **M. M. Taqui Khan and M. S. Jyoti,** *J. Inorg. Nucl. Chem.,* 1978, *40,* 1731.
144. **M. M. Taqui Khan and S. Satyanarayana,** *Indian J. Chem. Sect. A,* 1983, *22A,* 584.
145. **P. R. Reddy and B. M. Reddy,** *Polyhedron,* 1986, *5,* 1947.
146a. **J. A. Carrabine and M. Sundaralingam,** *J. Chem. Soc. Chem. Commun.,* 1968, 746.
146b. **M. Sundaralingam and J. A. Carrabine,** *J. Mol. Biol.,* 1971, *61,* 287.
147. **K. Saito, R. Terashima, T. Sakaki, and K. Tomita,** *Biochem. Biophys. Res. Commun.,* 1974, *61,* 83.
148. **T. J. Kistenmacher, D. Szalda, and L. G. Marzilli,** *Acta Crystallogr. Sect. B,* 1975, *B31,* 2416.
149. **D. J. Szalda, L. G. Marzilli, and T. J. Kistenmacher,** *Inorg. Chem.,* 1975, *14,* 2076.
150. **C. M. Mikulski, C. J. Lee, T. B. Tran, and N. M. Karayannis,** *Inorg. Chim. Acta,* 1987, *136,* L13.
151. **B. T. Khan and A. Mehmood,** *J. Inorg. Nucl. Chem.,* 1978, *40,* 1938.
152. **F. Coletta, R. Ettore, and A. Gambaro,** *J. Magn. Reson.,* 1976, *22,* 453.
153. **P.-C. Kong and F. D. Rochon,** *Can. J. Chem.,* 1981, *59,* 3293.
154. **S. Jaworski, H. Schöllhorn, P. Eisenmann, U. Thewalt, and B. Lippert,** *Inorg. Chim. Acta,* 1988, *153,* 31.
155. **K. P. Beaumont, C. A. McAuliffe, and M. E. Friedman,** *Inorg. Chim. Acta,* 1977, *25,* 241.
156. **M. M. Singh, Y. Rosopulos, and W. Beck,** *Chem. Ber.,* 1983, *116,* 1364.
157a. **L. G. Marzilli, T. J. Kistenmacher and M. Rossi,** *J. Am. Chem. Soc.,* 1977, *99,* 2797.
157b. **T. J. Kistenmacher, M. Rossi, and L. G. Marzilli,** *Inorg. Chem.,* 1979, *18,* 240.
158. **M. S. Holowczak, M. D. Stancl, and G. B. Wong,** *J. Am. Chem. Soc.,* 1985, *107,* 5789.
159. **C. Gagnon, A. L. Beauchamp, and T. Tranqui,** *Can. J. Chem.,* 1979, *57,* 1372.
160. **M. Authier-Martin and A. L. Beauchamp,** *Can. J. Chem.,* 1977, *55,* 1213.
161. **L. Prizant, R. Rivest, and A. L. Beauchamp,** *Can. J. Chem.,* 1981, *59,* 2290.
162. **J.-P. Charland, M. Simard, and A. L. Beauchamp,** *Inorg. Chim. Acta,* 1983, *80,* L57.
163. **E. Sinn, C. M. Flynn, Jr., and R. B. Martin,** *Inorg. Chem.,* 1977, *16,* 2403.
164. **B. Lippert, C. J. L. Lock, and R. A. Speranzini,** *Inorg. Chem.,* 1981, *20,* 335.
165. **H. Schöllhorn, U. Thewalt, G. Raudaschl-Sieber, and B. Lippert,** *Inorg. Chim. Acta,* 1986, *124,* 207.
166. **J. F. Britten, B. Lippert, C. J. L. Lock, and P. Pilon,** *Inorg. Chem.,* 1982, *21,* 1936.
167. **R. Beyerle-Pfnür, B. Brown, R. Faggiani, B. Lippert, and C. J. L. Lock,** *Inorg. Chem.,* 1985, *24,* 4001.
168. **R. Faggiani, B. Lippert, C. J. L. Lock, and R. A. Speranzini,** *Inorg. Chem.,* 1982, *21,* 3216.
169. **B. Lippert,** *J. Am. Chem. Soc.,* 1981, *103,* 5691.
170. **R. Faggiani, C. J. L. Lock, and B. Lippert,** *J. Am. Chem. Soc.,* 1980, *102,* 5418.
171. **R. Faggiani, B. Lippert, and C. J. L. Lock,** *Inorg. Chem.,* 1982, *21,* 3210.
172. **J. D. Orbell, L. G. Marzilli, and T. J. Kistenmacher,** *J. Am. Chem. Soc.,* 1981, *103,* 5126.
173. **A. P. Hitchcock, C. J. L. Lock, W. M. C. Pratt, and B. Lippert,** in *Platinum, Gold and Other Metal Chemotherapeutic Agents* (ACS Symp. Ser. 209), S. J. Lippard, Ed., American Chemical Society, Washington, D.C., 1983, 209.
174. **B. Lippert, C. J. L. Lock, and R. A. Speranzini,** *Inorg. Chem.,* 1981, *20,* 808.
175. **I. A. G. Roos, A. J. Thomson, and J. Eagles,** *Chem. Biol. Interact.,* 1974, *8,* 421.
176. **C. J. L. Lock, R. A. Speranzini, and J. Powell,** *Can. J. Chem.,* 1976, *54,* 53.
177. **B. E. Brown and C. J. L. Lock,** *Acta Crystallogr. Sect. C,* 1988, *C44,* 611.

178. **R. Faggiani, C. J. L. Lock, and B. Lippert,** *Inorg. Chim. Acta*, 1985, *106*, 75.
179. **B. Lippert,** *Inorg. Chim. Acta,* 1981, *56*, L23.
180. **R. Faggiani, B. Lippert, C. J. L. Lock, and R. A. Speranzini**, *J. Am. Chem. Soc.*, 1981, *103,* 1111.
181. **F. Schwarz, H. Schöllhorn, U. Thewalt, and B. Lippert,** in *Platinum and Other Metal Coordination Compounds in Cancer Chemotherapy,* M. Nicolini, Ed., Martinus Nijhoff, Boston, 1988, 728.
182. **H. Schöllhorn, R. Beyerle-Pfnür, U. Thewalt, and B. Lippert,** *J. Am. Chem. Soc.,* 1986, *108*, 3680.
183. **B. Lippert, H. Schöllhorn, and U. Thewalt,** *J. Am. Chem. Soc.*, 1986, *108*, 6616.
184. **R. Beyerle-Pfnür, H. Schöllhorn, U. Thewalt, and B. Lippert,** *J. Chem. Soc. Chem. Commun.,* 1985, 1510.
185. **B. J. Graves and D. J. Hodgson,** *J. Am. Chem. Soc.*, 1979, *101*, 5608.
186. **A. L. Beauchamp,** *Inorg. Chim. Acta,* 1984, *91,* 33.
187. **R. Weiss and H. Venner,** *Hoppe-Seyler's Z. Physiol. Chem.,* 1969, *350,* 1188.

Section 2

COMPLEXES INVOLVING NUCLEOSIDES OF THE PYRIMIDINE BASES

Badar Taqui Khan

INTRODUCTION

The pyrimidine nucleosides include cytidine, thymidine, and uridine. Of these, cytidine and thymidine are the constituents of the DNA strand and cytidine and uridine are part of the RNA strand. Of the three nucleosides mentioned, cytidine is the most thoroughly studied because of its importance in anticancer research.[82] It is the nucleoside which readily forms complexes with platinum(II) and other transition metal ions. Cytidine exists predominantly in the keto form with an ambidentate binding nature. It can form complexes with heterocyclic N(3) and the exocyclic C(4)–NH_2 and the C(2)=O groups.

The donor ability of various sites in nucleosides in general depends on

1. The relative basicity of the donor groups
2. Weak repulsive or attractive interaction between the exocyclic group on the nucleoside and the coordinated ligands on the metal complex
3. Steric interaction between the exocyclic group and other ligands on the metal complex

Since electron density is delocalized over the entire pyrimidine ring, it is generally difficult to predict as to which site will be the binding site for the metal complex. However, the principal site of coordination of cytidine to a metal ion is mostly the heterocyclic nitrogen atom N(3). The pK_a value for the N(3)–H group of cytidine is 4.2,[23] which is almost the same as that of cytosine. The binding of the metal ion via the amino group is difficult since the pK_a value of NH_2 is 16.0.[2] Mercury(II), however, can combine[20] with the nitrogen atom of the NH_2 group without dissociation of the proton. The least preferential coordination site in cytidine is the exocyclic C(2)=O group because of its poor donor capacity.

Thymidine and uridine are weaker donors than cytidine. The nucleosides exist in the keto form. The pK_a value for N(2)–H dissociation in thymidine[83] and uridine[70] are 9.7 and 9.2, respectively, which are about the same as in thymine and uracil. These nucleosides are mostly coordinated to the metal ion via N(3) atom.

The exocyclic C(2)–O and C(4)–O groups are very weak donors compared to the N(3)–H group and are not usually coordinated to a metal ion. In some cases, however, a heavy metal ion[21] such as Hg(II) is coordinated to C(4)=O.

Potentiometry is a useful technique for the determination of the association constants of the nucleosides with a proton or a metal ion. Most of the investigations covered in the section deal with potentiometry where the method was used to determine the composition of a binary or a ternary complex with the nucleosides and their corresponding stability constants. The proton and ^{13}C magnetic resonance spectroscopy are the most important techniques for determining the site of coordination of the metal ion to the ligands. The proton nuclear magnetic resonance spectrum of cytidine[20] gives two doublets for the C(5)–5H and C(6)–H protons at 5.60 and 7.35 ppm, respectively, with a spin-spin coupling, ^{J}H–H of about 7 Hz. The NH_2 protons are observed as a broad signal at 7.1 ppm.[20] In the ^{13}C nuclear magnetic resonance spectrum of cytidine the C(2), C(4), C(5), and C(6) carbon atoms are exhibited at 155.3, 160.3, 93.5, and 141.3 ppm, respectively. On coordination of a metal ion to the N(3) site, the C(5)–H and C(6)–H protons undergo a downfield shift of about 0.10 to 0.20 ppm with a greater shift of C(5)–H proton as compared to C(6)–H. The coordination of a metal ion to the N(3) site brings about a greater downfield[20] shift of C(2) and C(4) resonance (2 to 3 ppm) as compared to the C(5) and C(6) (1 to 2 ppm). In the mercury complex of cytidine the site of coordination to the metal ion, C(4)–NH_2 was established[20] on the basis of the downfield shift on the NH_2 signal by about 0.3 ppm. In this case, the C(4) carbon atoms undergo a maximum downfield shift of about 3 ppm as compared[20] to the other carbon atoms.

The proton nuclear magnetic resonance spectrum of thymidine[20] gives a singlet for the C_6–H proton at 7.1 ppm. This signal is shifted downfield by about 0.1 ppm on coordination to a metal ion. The proton magnetic resonance spectrum of uridine gives two doublets[20] corresponding to C(5)–H and C(6)–H protons at 5.36 and 7.88 ppm, respectively. These doublets are shifted downfield by a small magnitude on coordination of a metal to the N(3) site of the ligand. The C(2), C(4), C(5) and C(6) signals are shifted downfield by about 3 to 4 ppm as compared to the C(5) and C(6) signals which in certain cases are shifted upfield.

The lowering of the characteristic infrared frequencies on coordination is also used as a criterion of bonding of a particular site to the metal ion. In the case of cytidine, the lowering of the ring C=N and C=C stretching frequencies observed at 1480 and 1580 cm^{-1} in the ligand is considered a support for the involvement of ring nitrogen coordination to the metal ion. The NH_2 stretching in the ligand is observed at 3300 cm^{-1}. On coordination to mercury(II), this frequency[20] is lowered by about 100 cm^{-1}. The C(2)=O stretching frequency at 1710 cm^{-1} in the ligand is not affected[84] on coordination of the metal ion to the N(3) or the NH_2 site of the ligand. In uridine and thymidine the C(2)=O and the C(4)=O stretching frequency in the range 1640 to 1700 cm^{-1} is not affected[84] by complexation of the ligand through the N(3) position of the metal ion. The C=N and C=C ring stretching frequencies observed in the range 1390 to 1420 cm^{-1} and 1460 to 1480 cm^{-1} in uridine and thymidine are lowered to the extent of 60 to 70 cm^{-1} on complexation[20] of the metal ion to the N(3) position.

TABLE INDEX

Base no.	Compound	Base
C1n	Cytidine	cyd
C2n	*O*-Methoxycytidine	Omeocyd
C3n	2′-Deoxycytidine	dcyd
T1n	Thymidine	thd
T2n	1-Methylthymidine	1methd
T3n	2′-Deoxythymidine	dthd
U1n	Uridine	urd
U2n	2′-Deoxyuridine	2durd

TABLE 1

Base no.	Base	Metal	Stoichiometry	Method	Ref.
			Cytidine		
C1n	Cyd	Ag(I)	[Ag(cyd)]	nmr, ir	60
			Ag...cyd	nmr	60
		Au(III)	$[Au(cyd)_2Cl_2]Cl$	ir, nmr	17
				esr, ir	18
		Cd(II)	Cu...cyd	nmr	3
		Cu(II)	Cu...cyd	pot	4
			$[Cu(ala)_3(cyd)](ClO_4)_2$	nmr	5
			Cu...cyd	esr	6
			Cu...cyd	nmr	7
			$[Cu(cyd)_4(H_2O)_n]ClO_4$	X-ray	8
			Cu...cyd	spec, mag	9
			Cu...cyd...gly	esr, ks	10
			Cu...cyd...leu		
			Cu...cyd...trp		
			Cu...cyd...glygly	esr	12
			Cu...cyd	esr, nmr, spec	13
			Cu...cyd	nmr, spec	14
			Cu[(cyd)(OH)]Cl	esr, ir, spec	15
		Co(II)	$[Co(ala)_2(cyd)(MeOH]Cl_2$	nmr	5
			Co...cyd	pot	4
		Co(III)	Co...cyd	pot	16
		Hg(II)	Hg...cyd	nmr	20
			Hg...cyd	ir, nmr	2
			CH_3Hg...cyd	spec	21
		Mn(II)	Mn...cyd	pot	3
			Mn...cyd	nmr	19
		Ni(II)	Ni...cyd	pot	3
			$[Ni(ala)(cyd)Cl_2]$	nmr, ir	4
			(Ni...cyd...gua)	nmr	23
		Os(VI)	Os...cyd	kin	22
		Pd(II)	$[Pd(cyd)_4]Cl_2$	spec, nmr	24
			$[Pd(cyd)_4](PF_6)_2$	spec, nmr	24
			$K[Pd(cyd)Cl_3]$	ir, nmr	25
			cis$[Pd(guo)_2(cyd)_2]Cl_2$	ir, nmr	26
			trans$[Pd(guo)_2(cyd)_2]Cl_2$	ir, nmr	26
			cis$[Pd(cyd)_2Cl_2]$	ir, nmr	26
			$[Pd(cyd)_4]Cl_2$	ir, nmr	26
			$[Pd(dien)(cyd)]Cl_2$	kin, nmr	27
			$[(\eta^3\text{-}C_3H_5)(cyd)PdCl]$	ir, nmr, spec	28
			[(met)(cyd)PdCl]	ir, nmr, spec	28
			Pd...cyd	nmr	29
			$[(\eta^3\text{-}C_3H_5)(cyd)PdCl]$	nmr	30
			$[Pd(NH_3)_2(cyd)_2]Cl_2$	nmr	31
			cis & *trans*$[Pd(cyd)_2(glygly)]$	nmr	32, 33
			cis$[Pd(cyd)_2(mit)]Cl$	nmr	24
			cis & *trans*$[Pd(cyd)_2Cl_2]$	nmr, ir	34

TABLE 1 (CONTINUED)

Base no.	Base	Metal	Stoichiometry	Method	Ref.
			$[Pd(cyd)_2Cl_2]$	nmr	35
			$[Pd(cyd)Br_2]$	nmr	35
			$[Pd(dien)(cyd)](ClO_4)_2$	nmr	36
			$Pd(cyd)_2Cl_2$	nmr	37
			$Pd(cyd)Cl_2$	nmr	37
			Pd(dien)(cyd)	nmr	38
			$[Pd(cyd)Cl_2]$	nmr	38
			$[Pt(mit)(cyd)_2]Cl_2$	ir, nmr, spec	24
			$[Pt(mit)_2(cyd)_2]Br_2$	ir, nmr, spec	24
			$[Pt(NH_3)_2(cyd)Cl]Cl$	ir, nmr, spec	39, 40
			trans$[Pt(DMSO)(cyd)Cl_2]$	X-ray	41
			cis$[Pt(NH_3)_2(cyd)Cl_2]$	kin	42
			[Pt(dien)(cyd)Cl]Cl	nmr	43
			cis$[Pt(NH_3)_2(cyd)]^{2+}$	nmr	44
			$K[Pt(cyd)Cl(OH)]H_2O$	nmr, spec	45
			cis$[Pt(NH_3)_2Cl_2]$...cyd	nmr, spec	46
			$[Pt(cyd)_4]Cl_2$	nmr	47
			cis$[Pt(gly)(cyd)]Cl_2.2H_2O$	ir, nmr, spec	48
			cis$[Pt(ala)(cyd)]Cl_2.2H_2O$	ir, nmr, spec	48
			K[Ptme(COD)(cyd)]	ir, nmr, spec	49
			$H[PtCl_3(cyd)]$	X-ray, XPS	50
			cis$[Pt(NH_3)_2Cl_2]$ + cyd	nmr	51
			cis$[Pt(NH_3)_2Cl_2]$ + cyd	nmr	52
			cis & *trans*$[Pt(NH_3)_2Cl_2]$ + cyd	nmr	53
		Rh(I)	$[Rh(cyd)_3Cl]$	ir, nmr	54
			$[Rh(PPh_3)(CO)(cyd)]^+$	nmr, ir	55
			$[Rh(CO)_2(cyd)Cl]$	nmr, ir	56
		Rh(III)	$[Rh(cyd)_3Cl_3]$	ir, nmr	54
		Ru(III)	$[Ru(cyd)_3(CH_3OH_3](ClOH_4)$ $3CH_3O$	nmr, ir	57
			$[Ru(NH_3)_5(cyd)]Cl_3$	ir, spec	58
			$[Ru(cyd)_5(H_2O)](ClO_4)_3.2H_2O$	ir, nmr, spec, cond.	57
			$[Ru(NH_3)_5(cyd)]Cl_3$	spec	59
			$[Ru(H_2O)_5(cyd)](ClO_4)_3$	spec	57
			$Ru(alaH)_2(cyd)(MeOH)]Cl_3$	esr, ir, nmr	4
		Zn(II)	$[Zn(NO_3)_2(cyd)]$	ram, nmr	61
			Zn...cyd	ir, nmr	2
			Zn...cyd	pot	4
			***O*-Methoxycytidine**		
C2n	Omeocyd		$[Pt(Omeocyd)Cl]_2$	kin	62
			2′-Deoxycytidine		
C3n	dcyd		Cu...dcyd	esr	11

TABLE 1 (CONTINUED)

Base no.	Base	Metal	Stoichiometry	Method	Ref.
			Thymidine		
T1n	thd	Cd(II)	Cd...thd	ir	63
		Cu(II)	Cu...thd	esr, ir	13
		Hg(II)	Hg...thd	ir	1
			(PhHg)...thd	X-ray	64
		Mn(II)	Mn...thd	ir	63
		Os(VI)	Os...thd	kin, ir	22
		Pd(II)	$[Pd(en)(OH)_2(urd)]$	ks	65
			Pd...urd	nmr	29
			Pd...urd	X-ray	66
		Pt(II)	$[Pt(bdppe)_2(dmf)(thd)]$	nmr, kin	67
			Pt...thd	X-ray	66
			Pt...thd	nmr, ir, spec	68
			$[Pt(NH_3)_2Cl_2]$...thd	nmr	52
		Rh(I)	$[(PPh_3)_2Rh(CO)]^+$...thd	nmr	55
		Ru(III)	$[Ru(thd)_2(H_2O)_4](ClO_4)_3$	nmr, ir	57
			$[Ru(thd)_2(H_2O)_2ClO_4](ClO_4)_2$	nmr, ir	57
			$[Ru(NH_3)_5(thd)]Cl_3$	spec, ir	15
			1-Methylthymidine		
T2n	1methd	Hg(II)	Hg...1methd	ram	47
			2′-Deoxythymidine		
T3n	dthd	Cu(II)	Cu...dthd	esr	9
			Uridine		
U1n	Urd	Ag(I)	Ag...urd	X-ray	35
		Cd(II)	Cd...urd	X-ray	63
		Cr(III)	Cr...urd	ir, spec	69
		Cu(II)	Cu...urd	pot	70
			Cu...urd	nmr	7
			Cu...urd	esr, nmr, ir	13
			$[Cu(urd)(ClO_4)_2](H_2O)_2$	X-ray	8
			Cu...urd	kin	71
			Cu...urd	esr	9
			$[Cu_2(ac)_4]$ + urd	esr, spec, MD	72
			Cu...urd...gly	esr, ks	10
			Cu...urd...leu	esr, ks	10
			Cu...urd...trp	esr, ks	10
			Cu...urd	esr, ks	12
			Cu...urd	esr	6
			Cu...urd	esr	73
			Cu...urd	esr, ir, ram	13
		Co(II)	Co...urd	pot	2, 63

TABLE 1 (CONTINUED)

Base no.	Base	Metal	Stoichiometry	Method	Ref.
			Co...urd	ks	16
		Hg(II)	Hg...urd	ram, nmr	2
			meHg...urd	ram	21
			PhHg...urd	X-ray	64
		Mn(II)	Mn...urd	nmr	74
		Ni(II)	Ni...urd	pot	69
		Os(VI)	Os...urd	spec, kin	22
		Pd(II)	$[Cl_2(PBu_3)Pd(urd)Pd(PBu_3)Cl]_2$	X-ray	75
			$[Pd(en)(OH_2)(urd)]^{2+}$	ks	65
			Pd...urd	ks	66
			Halo-urd	X-ray	76
			Meo-urd	X-ray	76
		Pt(II)	$[PtCl_2(urd)gly]$	pot, nmr	48
			$[PtCl_2(urd)ala]$	pot, nmr	48
			cis$[Pt(NH_3)_2(urd)]$	nmr	77
			$[Pt(bdppe)_2(dmf)(urd)]$	nmr, kin	67
			Pt...urd	X-ray	66
			Pt...urd	ram, nmr	2
			cis$[Pt(cpa)_2urd](Ph_3BCN)_{1.25}$	X-ray	
			cis$[Pt(NH_3)_2urd](Ph_3BCN)_{1.25}$	X-ray	78
			$[Pt(NH_3)_3urd]$	X-ray	79
			cis$[Pt(NH_3)_2Cl_2]$ + urd	nmr	46
		Rh(I)	$[Rh(PPh_3)_2(CO)]^+$ + urd	nmr	55
		Rh(II)	Rh...urd	ram, nmr	2
		Rh(III)	$[Ru(NH_3)_2(urd)]Cl_3$	spec, ir	58
		Ru(III)	$[Ru(urd)(H_2O)_3(ClO_4)_2]ClO_4$	ir, nmr	57
			2′-Deoxyuridine		
U2n	durd	Cu(II)	Cu...durd	ks	80
			Cu...durd	ks	81

REFERENCES

1. **T. Yukono, S. Shimokawa, H. Fukumi, and J. Sohma,** *Nippon Kagaku Zaishi,* 1973, *2,* 201.
2. **L. G. Marzilli, B. de Castro, J. P. Caradonna, R. C. Stewart, and C. P. Van Vuuren,** *J. Am. Chem. Soc.*, 1980, *102*, 916.
3. **C. R. Krishnamurthy, R. Van Eldik, and G. M. Harris,** *J. Coord. Chem.*, 1980, *10,* 195.
4. **B. T. Khan and R. M. Raju,** *Indian J. Chem. Sect. A*, 1981, *20,* 680.
5. **B. T. Khan and M. Ali,** unpublished.
6. **Y.-Y. H. Chao and D. R. Kearns,** *J. Am. Chem. Soc.*, 1977, *99*, 6425.
7. **G. Kotowycz,** *Can. J. Chem.*, 1976, *52,* 924.
8. **J. B. Jean, J. Yves, and M. Alian,** *C. R. Acad. Sci. Ser. C.*, 1975, *280,* 827.
9. **K. Maskos,** *Acta Biochim. Pol.*, 1974, *21*, 255.
10. **K. Maskos,** *Acta Biochim. Pol.*, 1985, *32*; *Chem. Abstr.*, 1985, *104*, 47413t.
11. **H. Fritzsche, D. Tresselt, and Ch. Zimmer,** *Experienta*, 1971, *27*, 1253.
12. **S. V. Deshpande, R. K. Sharma, and T. S. Srivastava,** *Inorg. Chim. Acta,* 1983, *78*, 13.
13. **K. Maskos,** *Acta Biochim. Pol.*, 1979, *26*, 249.
14. **J. Dehand, J. Jordanov, and F. Keck,** *Inorg. Chim. Acta,* 1977, *21*, L13; *Chem. Abstr.*, 1977, *86,* 114709u.
15. **D. C. H. Nelson and F. J. Villa,** *J. Inorg. Nucl. Chem.*, 1980, *42,* 1669.
16. **T. Sorrel, L. A. Epps, T. J. Kistenmacher, and L. G. Marzilli,** *J. Am. Chem. Soc.*, 1977, *99,* 2173.
17. **N. Hadjiliadis, G. Pneumatikakis, and R. Basosis,** *J. Inorg. Biochem.,* 1981, *14,* 115; *Chem. Abstr.*, 1981, *95*, 81383p.
18. **D. Chatterji, U. S. Nandi, and S. K. Podder,** *Biopolymers,* 1977, *16,* 1863; *Chem. Abstr.*, 1977, *87,* 162948p.
19. **H. Fritzsche, K. Arnold, and R. Krusche,** *Stud. Biophys.*, 1974, *45*, 131; *Chem. Abstr.*, 1975, *82,* 125554h.
20. **B. T. Khan,** *Proc. Indian Natl. Sci. Acad.*, 1988, *55,* 446.
21. **S. Mansy, T. E. Wood, J. C. Sprowles, and R. S. Tobias,** *J. Am. Chem. Soc.*, 1974, *96,* 1762.
22. **F. B. Daniel and E. J. Behrman,** *J. Am. Chem. Soc.*, 1975, *97*, 7352.
23. **B. T. Khan, R. M. Raju, and S. M. Zakeeruddin,** *J. Chem. Soc. Dalton Trans.*, 1988, *(1),* 67.
24. **J. Dehand and J. Jordanov,** *J. Chem. Soc. Dalton Trans.*, 1977, 1588.
25. **N. Hadjiliadis and G. Pneumatikakis,** *J. Chem. Soc. Dalton Trans.*, 1978, 1691.
26. **G. Pneumatikakis, N. Hadjiliadis, and T. Theophanides,** *Inorg. Chem.*, 1978, *17*, 915.
27. **R. Menard, M. T. Phan Viet, and M. Zoder,** *Inorg. Chim. Acta,* 1987, *136*, 25.
28. **B. T. Khan and K. M. Mohan,** Unpublished.
29. **D. J. Nelson, P. L. Yeagle, T. L. Miller, and R. B. Martin,** *Bioinorg. Chem.*, 1976, *5,* 353.
30. **Y. Rosopulos, U. Negal, and W. Beck,** *Chem. Ber.*, 1985, *118,* 931.
31. **G. Pneumatikakis,** *Inorg. Chim. Acta,* 1982, *66*, 131.
32. **B. Jezowska-Trzebiatowska and S. Wolowiec,** *Biochim. Biophys. Acta,* 1982, *708, (1),* 12.
33. **B. Jezowska-Trzebiatowska and S. Wolowiec,** *Acta Biochim. Pol.*, 1983, *30*, 277; *Chem. Abstr.*, 1984, *101, (3-4),* 67944h.
34. **G. Pneumatikakis, N. Hadjiliadis, and T. Theophanides,** *Inorg. Chem.,* 1978, *17*, 915.
35. **P. C. Kong and F. D. Rochon,** *Can. J. Chem.*, 1981, *59,* 3293; *Chem. Abstr.*, 1981, *95,* 231101m.
36. **F. D. Rochon, P. Kong, B. Coulombe, and R. Melanson,** *Can. J. Chem.*, 1980, *58,* 381; *Chem. Abstr.*, 1980, *92*, 190541x.
37. **R. Ettore,** *Inorg. Chim. Acta,* 1978, *80*, L309; *Chem. Abstr.*, 1978, *90,* 13008s.

38. **E. Matczak-Jon, B. Jezowska-Trzebiatowska, and H. Kozlowski,** *J. Inorg. Biochem.*, 1980, *12,* 143; *Chem. Abstr.*, 1980, *93,* 39773z.
39. **A. I. Stetsenko, L. B. Selderkhanova, and A. I. Mokhov,** *Koord. Khim.*, 1985, *11,* 816.
40. **A. I. Stetsenko, G. M. Alekseeva, and K. I. Yakovlev,** *Zh. Neorg. Khim.*, 1985, *30,* 2592.
41. **R. Melanson and F. D. Rochon,** *Inorg. Chem.*, 1978, *17,* 679.
42. **W. M. Scovell and T. O'Connor,** *J. Am. Chem. Soc.*, 1977, *99,* 120.
43. **P. D. Kaplan, F. Smidt, A. Brause, and M. Orchin,** *J. Am. Chem. Soc.*, 1969, *91,* 85.
44. **G. Y. H. Chu, R. E. Duncan, and R. S. Tobias,** *Inorg. Chem.*, 1977, *16,* 2625.
45. **R. Ettore,** *Inorg. Chim. Acta*, 1980, *46,* L27; *Chem. Abstr.*, 1980, *92,* 208195z.
46. **J. P. Laurent and P. Lepage,** *Can. J. Chem.*, 1981, *59,* 1083; *Chem. Abstr.*, 1981, *95,* 43562p.
47. **S. Mansy and R. S. Tobias,** *Inorg. Chem.*, 1975, *14,* 287.
48. **B. T. Khan, G. N. Goud, and S. V. Kumari,** *Inorg. Chim. Acta,* 1983, *80,* 145.
49. **S. Komiya, Y. Mizuno, and T. Shibuya,** *Chem. Lett.*, 1986, *7,* 1065.
50. **P. Umapathy, R. A. Harnesswala, and C. S. Dorai,** *Polyhedron,* 1985, *4 (9),* 1595.
51. **M. W. K. Nee and J. D. Roberts,** *Biochemistry*, 1982, *21,* 4920; *Chem. Abstr.*, 1982, *97,* 138224k.
52. **W. Tang, S. Zhang, C. Yuan, and A. Tai,** *Gaodeng Xuexiao Huaxue Xuebao*, 1984, *5 (1),* 1; *Chem. Abstr.*, 1984, *100,* 202374u.
53. **S. Mansy, B. Rosenberg, and A. J. Thomson,** *J. Am. Chem. Soc.*, 1973, *95,* 1633.
54. **G. Pneumatikakis, J. Markopoulos, and A. Yannopoulos,** *Inorg. Chim. Acta,* 1987, *136,* L25; *Chem. Abstr.*, 1987, *107,* 88454n.
55. **D. W. Abbott and C. Woods,** *Inorg. Chem.*, 1983, *22,* 2918.
56. **M. M. Singh, Y. Rosopulos, and W. Beck,** *Chem. Ber.*, 1983, *116,* 1364; *Chem. Abstr.*, 1983, *99,* 81480a.
57. **B. T. Khan, A. Gaffuri, P. N. Rao, and S. M. Zakeeruddin,** *Polyhedron,* 1987, *6 (3),* 387.
58. **B. T. Khan, A. Gaffuri, and M. R. Somayajulu,** *Indian J. Chem. Sect. A,* 1981, *20,* 189.
59. **M. J. Clarke,** *J. Am. Chem. Soc.*, 1978, *100,* 5068.
60. **R. Cini, P. Colamarino, and P. L. Orioli,** *Bioinorg. Chem.*, 1977, *7,* 345; *Chem. Abstr.*, 1977, *88,* 114514t.
61. **L. G. Marzilli, R. C. Stewart, C. P. Van Vuuren, B. de Castro, and J. P. Caradonna,** *J. Am. Chem. Soc.*, 1978, *100,* 3967.
62. **N. Hadjiliadis and G. Pneumatikakis,** *Inorg. Chim. Acta,* 1980, *46,* 255; *Chem. Abstr.*, 1980, *94,* 209165t.
63. **M. Goodgame and W. K. Johns,** *J. Chem. Soc. Dalton Trans.*, 1978, 1294.
64. **P. Peringer,** *Z. Naturforsch. Teil B,* 1979, *34(B),* 1459; *Chem. Abstr.*, 1979, *92,* 94670t.
65. **M. C. Lim,** *J. Inorg. Nucl. Chem.*, 1981, *43,* 221; *Chem. Abstr.*, 1981, *94,* 146074.
66. **M. C. Lim and R. B. Martin,** *J. Inorg. Nucl. Chem.*, 1976, *38,* 1915; *Chem. Abstr.*, 1976, *86,* 47720a.
67. **B. Longato, B. Corain, G. H. Bonora, and G. Pilloni,** *Inorg. Chim. Acta,* 1987, *137,* 75.
68. **K. Inagaki and Y. Kidani,** *Bioinorg. Chem.*, 1978, *9,* 333; *Chem. Abstr.*, 1978, *90,* 82362q.
69. **J. J. Fiol, A. Terron, and V. Moreno,** *Polyhedron,* 1986, *5,* 1125.
70. **B. T. Khan, R. M. Raju, and S. M. Zakeeruddin,** *J. Coord. Chem.*, 1988, *16(3), 237.*
71. **I. Sovago and R. B. Martin,** *Inorg. Chim. Acta*, 1980, *46,* 91.
72. **P. Chalilpoyil and L. G. Marzilli,** *Inorg. Chem.*, 1979, *18,* 2328.
73. **N. A. Berger and G. L. Eichhorn,** *Biochemistry*, 1971, *10,* 1857.
74. **G. Kotowycz and O. Suzuki,** *Biochemistry,* 1973, *12,* 3434.
75. **M. W. Beck, J. C. Calarese, and N. D. Kottmair,** *Inorg. Chem.*, 1979, *18,* 176.
76. **D. E. Bergstrom and K. M. Ogawa,** *J. Am. Chem. Soc.*, 1978, *100,* 8106.
77. **G. Y. H. Chu, R. E. Duncan, and R. S. Tobias,** *Inorg. Chem.*, 1977, *16,* 2625.
78. **T. Boon-Keng, K. Kijima, and R. Bau,** *J. Am. Chem. Soc.*, 1978, *100,* 621.
79. **K. Inagaki and Y. Kidani,** *Bioinorg. Chem.*, 1978, *9,* 157.

80. **Y. L. Tan and A. Beck,** *Biochem. Biophys. Acta,* 1973, *299*, 500; *Chem. Abstr.*, 1973, *79,* 19007k.
81. **H. Fritzsche, D. Tresselt, and Ch. Zimmer,** *Experientia*, 1971, *27(11),* 1253.
82. **B. Lippert,** in *Platinum, Gold and Other Metal Chemotherapeutic Agents* (ACS Symp. Ser. 209), S. J. Lippard, Ed., Washington, D.C., 1983.
83a. **A. I. Stetsenko, K. I. Yakolev, and S. A. Dyachenko,** *Russ. Chem. Rev.,* 1987, *56*, 875.
83b. **K. Nakanishi, N. Suzuki, and F. Imaziki,** *Bull. Chem. Soc. Jpn.*, 1961, *34*, 53.
84. **B. Lippert, H. Schöllhorn, and W. Thewaltz,** *Z. Naturforsch.,* 1983, *38*, 1441.

Section 3

COMPLEXES INVOLVING NUCLEOTIDES AND OLIGONUCLEOTIDES OF THE PYRIMIDINE BASES

Badar Taqui Khan

INTRODUCTION

The monophosphates of cytidine, uridine, and thymidine can exist as the 3′ or 5′ isomers. It is interesting to note that the interaction of transition metal ions with the nucleotide monophosphates takes place with N(3) ring nitrogen;[7,11,13,24] the phosphate group is not involved in bonding. These results are based on crystallographic[11,13] investigation of Pt(II) complexes and the ^{13}C resonance spectroscopy of the nucleotide phosphates.[25] In the Mn(II) complex, however, the C(2)=O of 5′-CMP and C(2)=O and C(4)=O of 5'-UMP and 5′-TMP are involved in bonding with no involvement of the phosphate group. Potentiometric investigation[3] of 5′-CMP, 5′-TMP, and 5′-UTP complexes of Cu(II), Ni(II), Co(II), Cu(II), and Zn(II), however supports the coordination of the phosphate group to the metal ion.

Interaction of metal ions with 5′-cytidine triphosphate (CTP) and 5′-uridine triphosphate (UTP) was studied mostly by potentiometry[22] and ^{1}H, ^{13}C nuclear magnetic resonance spectroscopy.[5,20] In these complexes the metal ion is usually bonded to the phosphate group with the bases free to stalk in solution.[22]

Quantitative information on the interaction of metal ions with oligo- and polynucleotides is not extensive. Proton magnetic resonance studies on the Cu(II) interaction with the polyadenine-polyuridine double helix[20] indicate that the metal binding site is N(3) on poly U. Interaction of methyl mercury with poly U studied by Raman difference spectroscopy[24] also indicates N(3) as the site of coordination to the metal ion.

TABLE INDEX

Base no.	Compound	Base
C1na	Cytidine-5′-monophosphate	CMP
C1nb	Cytidine-3′-monophosphate	3CMP
C1nc	Cytidine-5′triphosphate	CTP
T1na	Thymidine-5′monophosphate	TMP
T1nb	Thymidine-5′-triphosphate	TTP
U1na	Uridine-5′-monophosphate	UMP
U1nb	Uridine-5′-triphosphate	UTP

TABLE 1

Base no.	Base	Metal	Stoichiometry	Method	Ref.
			Cytidine-5′-monophosphate		
C1na	CMP	Co(II)	Co...CMP	pot	6
			Co...CMP	pot	3
		Cr(III)	$[Cr(en)(CMP)_2]8H_2O$	ir, spec	5
		Cu(II)	[Cu(CMP)2]	mag, spec, esr	1
			Cu...CMP	esr, ir	2
			Cu...CMP	pot	3
			Cu...CMP	nmr	4
		Hg(II)	Hg...CMP	ram	9
		Mn(II)	Mn...CMP	pot	7
			Mn...CMP	pot	3
		Ni(II)	Ni...CMP	pot	3
			Ni...CMP	pot	6
			Ni...CMP	pot	10
		Pt(II)	$[Pt(NH_3)_2(CMP)_2]^{2-}$	X-ray	11
			$[Pt(bipy)(en)]^{2+}$...CMP	nmr	8
		Zn(II)	Zn...CMP	pot	6
			Zn...CMP	pot	3
			Zn...CMP	pot	12
			Cytidine-3′-monophosphate		
C1nb	3CMP	Pt(II)	$bis[Pt(NH_3)_2(3CMP)]_2$	X-ray	13
			Cytidine-5′-triphosphate		
C1nc	CTP	Co(II)	Co...CTP	pot	16
			Co...CTP	pot	17
		Cu(II)	Cu...CTP	pot	14
			Cu...CTP	pot	15
		Ni(II)	Ni...CTP	pot	16
			Ni...CTP	pot	10
		Zn(II)	Zn...CTP	pot	16
			Thymidine-5′-monophosphate		
T1na	TMP	Co(II)	Co...TMP	pot	3
		Cu(II)	Cu...TMP	nmr	4
			Cu...TMP	pot	3
			Cu...TMP	esr, ir	7
		Mn(II)	Mn...TMP	nmr	2
			Mn...TMP	pot	3
		Ni(II)	Ni...TMP	pot	3
		Zn(II)	Zn...TMP	pot	3

TABLE 1 (CONTINUED)

Base no.	Base	Metal	Stoichiometry	Method	Ref.
			Thymidine-5′-triphosphate		
T1nb	TTP	Cu(II)	Cu...TTP	pot	14
		Mn(II)	Mn...TTP	pot	18
		Zn(II)	Zn...TTP	pot	19
			Uridine-5′-monophosphate		
U1na	UMP	Co(II)	Co...UMP	pot	3
		Cr(III)	Cr...UMP	ir	21
			$[Cr(en)(UMP)(OH)]3H_2O$	ir, spec	5
		Cu(II)	Cu...UMP	nmr	4
			Cu...UMP	pot	3
			Cu...UMP	esr, ir	20
			Cu...UMP	nmr	20
		Mn(II)	Mn...UMP	nmr	5
			Mn...UMP	pot	3
			Mn...UMP	nmr	7
		Ni(II)	Ni...UMP	pot	3
		Pt(II)	$[Pt(bipy)(en)]^{2+}$...UMP	nmr	8
		Zn(II)	Zn...UMP	pot	3
			Uridine-5′-triphosphate		
U1nb	UTP	Cu(II)	Cu...(bipy)...UTP	pot	22
			Cu...(phen)...UTP	pot	22
			Cu...UTP	pot	14
			Cu...UTP	pot	23
		Co(II)	Co...UTP	pot	17
			Co...UTP	pot	23
		Mn(II)	Mn...UTP	pot	23
			Mn...UTP	pot	18
		Ni(II)	Ni...UTP	pot	23
		Zn(II)	Zn...UTP	pot	23
			Zn...UTP	pot	18

REFERENCES

1. **D. C. H. Nelson and F. J. Vella,** *J. Inorg. Nucl. Chem.,* 1980, *42,* 1669.
2. **K. Maskos,** *Acta Biochim. Pol.,* 1979, *26,* 249.
3. **S. S. Massoud and H. Sigel,** *Inorg. Chem.,* 1988, *27,* 1447.
4. **G. Kotowycz,** *Can. J. Chem.,* 1976, *52,* 924.
5. **A. M. Calafat, D. Mulet, J. J. Fiol, and A. Terron,** *Inorg. Chim. Acta,* 1987, *138,* 105.
6. **P. A. Monorik, N. K. Davidenko, N. P. Aleksyuk, and E. I. Lopatina,** *Zh. Neorg. Khim.,* 1984, *29,* 735.
7. **G. Kotowycz and O. Suzuki,** *Biochemistry,* 1973, *12,* 3434.
8. **O. Yamauchi, A. Odani, R. Shimata, and Y. Kosaka,** *Inorg. Chem.,* 1986, *25,* 3337.
9. **R. W. Chrisman, S. Mansy, H. J. Peresie, A. Ranade, T. A. Berg, and R. S. Tobias,** *Bioinorg. Chem.,* 1977, *7,* 245.
10. **C. M. Frey and J. E. Stuehr,** *J. Am. Chem. Soc.,* 1972, *94,* 8898.
11. **S. Louie and R. Bau,** *J. Am. Chem. Soc.,* 1977, *99,* 3874.
12. **G. Weitzel and T. Speer,** *Z. Physiol. Chem.,* 1958, *313,* 212.
13. **S.-M. Wu and R. Bau,** *Biochem. Biophys. Res. Commun.,* 1979, *88,* 1435.
14. **H. Sigel,** *Eur. J. Biochem.,* 1968, *3,* 530.
15. **H. Sigel, D. H. Buisson, and B. Prijs,** *Bioinorg. Chem.,* 1975, *5,* 1.
16. **M. M. T. Khan and P. R. Reddy,** *J. Inorg. Nucl. Chem.,*1975, *37,* 771.
17. **E. Walaas,** *Acta Chim. Scand.,* 1958, *12,* 528; 1957, *11,* 1002.
18. **H. Sigel,** *J. Am. Chem. Soc.,* 1975, *97,* 3209.
19. **H. Sigel,** *J. Inorg. Nucl. Chem.,* 1977, *39,* 1903.
20. **N. A. Berger and G. L. Eichhron,** *Biochemistry,* 1971, *10,* 1857.
21. **J. J. Fiol, A. Terron, and V. Moreno,** *Polyhedron,* 1986, *5,* 1125.
22. **R. Tribolet, R. Malini-Balakrishnan, and H. Sigel,** *J. Chem. Soc. Dalton Trans.,* 1985, 2291.
23. **M. M. T. Khan and P. R. Reddy,** *J. Inorg. Nucl. Chem.,* 1976, *38,* 1234.
24. **S. Mansy, T. E. Wood, J. C. Sprowles, and R. S. Tobias,** *J. Am. Chem. Soc.,* 1974, *96,* 1762.
25. **Y.-Y. H. Chao and D. R. Kearns,** *J. Am. Chem. Soc.,* 1977, *99,* 6425.

Section 4

COMPLEXES INVOLVING PURINE BASES AND THEIR DERIVATIVES

James R. Lusty, Peter Wearden, and Hardy S. O. Chan

INTRODUCTION

The interaction of nucleobases with transition metals has interested coordination chemists for several decades. This interest increased dramatically after the discovery that *cis*-diamminedichloroplatinum(II) was effective against certain types of cancer. The target of this interaction was shown to be guanosine residues in DNA. The interaction of metals with purine nucleosides (Section 5) and their nucleotides, oligonucleotides, and DNA (Section 6) is covered elsewhere in this publication, and arguments relating to the applications and significance of this work will not be rehearsed here. This section deals with the interaction of purine nucleobases and their derivatives with transition metals.

The interaction of the nucleotide, or even the nucleoside, poses several problems, as there are a considerable number of alternative bonding sites.

The acid-base properties of uracil were considered by Lippert[221] who formulated the possibilities involving mono- and di-metal coordination for platinum(II). In xanthine, a similar base to uracil, with two exocyclic carbonyl groups, we have predicted over 15 different species for single platinum coordination.[228] Many of the complex species for uracil and for some of the purine bases have been identified by nuclear magnetic resonance spectroscopy.

Although the best defined structures are given by X-ray investigation, solution studies, primarily by nuclear magnetic resonance, but also by infrared spectroscopy, provide information under biologically relevant conditions. There are obvious problems in the interpretation and comparison of the data. While solid studies can be definitive, particularly where several techniques are used, solution studies are often inconclusive, although the use of ^{195}Pt, ^{13}C, and ^{15}N nuclear magnetic resonance has increased the scope and reliability of the interpretation. Some solution studies which do not give definitive products are included in this section, but they relate to measured and recorded interactions between species.

There are a number of factors affecting the binding mode of the base, and these are reviewed in several articles.[138,139,147] In order to simplify the large number of interactions possible between nucleotides and transition metals, model compounds of nucleobase derivatives have been extensively used. In order to replicate the behavior, the N(9) position has to be blocked, and this is usually achieved using an alkyl substituent. Hence, in this section there are considerable numbers of complexes reported, involving 9-methyl- and 9-ethylpurines, especially of guanine and adenine.

While it is generally accepted that N(7) is the preferred coordination site in 6-oxopurine systems (and is believed to be the primary target for *cis*-diamminedichloroplatinum(II) as an antitumor drug) some controversy has existed over the possibility of a second linkage at C(6)=0.[56,141,165,263]

Most of the evidence to support such a linkage is based on infrared spectroscopy and, in particular, a lowering of the (C(6)=O) frequency from about 1700 to about 1620 cm^{-1}.[122,148] The existence of N(7)/O(6) chelation for a derivative of xanthine was demonstrated using X-ray and electron spin resonance techniques for the complex *bis* (π-cyclopentadienyl) (thiophyllinato) titanium (III),[62] and O(6) interaction has been implicated for a number of complexes where N(7) is also blocked. For example, in 7,9-dimethylhypoxanthine, some participation of O(6) in the metal binding scheme for Cu(II)[51] and Pt(II)[71,118] was reported. However, it is now generally accepted that while the N(7)/O(6) chelation is possible, monodentate binding of a nucleobase via N(7) is capable in itself of causing base mispairing, and the NH_3 ligands appear to contribute not only to the mispairing, but also to a lack of base selectivity.[154] Further, it has been argued that although the clip for N(7)/S(6) exists in 9-methyl-6-thiopurine[232] and 9-benzyl-6-thiopurine,[217] there is no definitive evidence for the clip with oxygen, even in anionic guanine complexes.[132] In addition, deprotonation at N(1) causes shifts of up to 70 cm^{-1} in the (C=O) frequency,[132] and polymeric formation, which occurs for several oxo-purines, using N(1) and N(7), could be responsible for the observed shifts in the infrared spectra.

Thiopurine complexes have been included in this section as they have been used as synthetic analogues for a number of naturally occurring bases; in particular, 6-thiopurine (as an analogue for hypoxanthine) and 6-thioguanine have been extensively studied. Much of the information currently available for the former is reviewed in a recent publication.[146]

There are many examples of complexes binding through alternative sites to those mentioned above. Methyl derivatives have been used to good effect as blocking agents. Caffeine (1,3,7-trimethylxanthine), theophylline (1,3-dimethylxanthine), and theobromine (3,7-dimethylxanthine) have all been used in a variety of solvents and conditions in order to produce complexes with a range of donor atoms.

The reader is referred to reviews on individual aspects of these interactions including binding site preferences,[138,139,147,191,192] methods of study,[147,191] platinum group metal interactions,[143] kinetics of interaction of *cis*-diamminedichloroplatinum(II),[200,240] and transition metal complexes in cancer chemotherapy.[140,144,147]

In addition, relevant conference proceedings[130,131,133,136] and the series *Advances in Inorganic Biochemistry*[138,142] and *Metal Ions in Living Systems*[139] give excellent coverage of these interactions.

TABLE INDEX

Base no.	Compound	Base
A1	Adenine	ade
A2	Adenine-*N*(1)-oxide	adelox
A3	8-Azaadenine	8azade
A4	3-Benzyladenine	3bzade
A5	3-((Ethoxycarbonyl)methyl)adenine	ecma
A6	3-Methyladenine	3meade
A7	9-Methyladenine	9meade
A8	1,9-Dimethyladenine	19dmeade
A9	2,9-Dimethyladenine	29dmeade
A10	8,9-Dimethyladenine	89dmeade
A11	3-(γ,γ-Dimethylallyl)adenine	tct
X30	Caffeine	caf
G1	Guanine	gua
G2	9-Ethylguanine	9etgua
G3	1-Methylguanine	1megua
G4	9-Methylguanine	9megua
G5	N(2),N(2)-Dimethyl-9-methylguanine	tmegua
G6	N(2),N(2)-Dimethyl-9-propylguanine	dmeprgua
G7	1,9-Dimethylguanine	19dmegua
G8	7,9-Dimethylguanine	79dmegua
G9	6-Selenoguanine	6Segua
G10	6-Thioguanine	6Sgua
G11	8-Thioguanine	8Sgua
H1	Hypoxanthine	hyp
H2	8-Azahypoxanthine	8azhyp
H3	1-Methylhypoxanthine	1mehyp
H4	7-Methylhypoxanthine	7mehyp
H5	7-Methyl-9-propylhypoxanthine	7me9prhyp
H6	9-Methylhypoxanthine	9mehyp
H7	7,9-Dimethylhypoxanthine	dmehyp
H8	1,4-Bis(hypoxanth-9-yl)butane	hypbu
H9	1,3-Bis(hypoxanth-9-yl)-2-propanol	hyppo
H10	1,3-Bis(hypoxanth-9-yl)propane	hyppr
P1	Purine	pur
P2	Purine-*N*(1)-oxide	pur1ox
P3	2-Amino-9-methylpurine	2A9mepur
P4	8-Amino-9-methylpurine	8A9mepur
P5	6-Aminobenzylpurine	6Abzpur
P6	6-Aminoethylaminepurine	6Aeapur
P7	2,6-Diaminopurine	26dApur
P8	2-Cloro-9-methylpurine	2Cl9mepur
P9	8-Chloro-9-methylpurine	8Cl9mepur
P10	6-Hydroxyethylaminepurine	6HOeapur
P11	6-Hydroxyethylmethylaminepurine	6HOemapur
P12	2-Methoxy-9-methylpurine	2meo9mepur

TABLE INDEX (CONTINUED)

Base no.	Compound	Base
P13	8-Methoxy-9-methylpurine	8meo9mepur
P14	9-Methylpurine	9mepur
P15	2,9-Dimethylpurine	29dmepur
P16	6,9-Dimethylpurine	69dmepur
P17	8,9-Dimethylpurine	89dmepur
P18	6-Thiopurine	6Spur
P19	9-Benzyl-6-thiopurine	9bz6Spur
P20	6-Butylthiopurine	6buSpur
P21	9-Methyl-6-thiopurine	9me6Spur
P22	1,9-Dimethyl-6-thiopurine	19dme6Spur
P23	2,9-Dimethyl-2-thiopurine	29dme2Spur
P24	2,9-dimethyl-8-thiopurine	89dme8Spur
X20	Theobromine	thb
X10	Theophylline	thp
X11	8-Ethyltheophylline	8etthp
X12	7-Methyltheophylline	7methp
X13	8-Pentyltheophylline	8pethp
X14	8-Phenyltheophylline	8phthp
X15	8-Propyltheophylline	8prthp
X16	8-Isopropyltheophylline	8iprthp
X1	Xanthine	xan
X2	8-Ethylxanthine	8etxan
X3	8-Ethyl-1-methylxanthine	8et1mexan
X4	8-Ethyl-3-methylxanthine	8et3mexan
X5	1-Methylxanthine	1mexan
X6	3-Methylxanthine	3mexan
X7	7-Methylxanthine	7mexan
X8	8-Methylxanthine	8mexan
X9	9-Methylxanthine	9mexan
X10	1,3-Dimethylxanthine	thp
X11	1,9-Dimethylxanthine	19dmexan
X20	3,7-Dimethylxanthine	thb
X25	3,8-Dimethylxanthine	38dmexan
X26	3,9-Dimethylxanthine	39dmexan
X30	1,3,7-Trimethylxanthine	caf
X35	1,3,8-Trimethylxanthine	8tmexan
X36	1,3,9-Trimethylxanthine	9tmexan
X37	2-Thioxanthine	2Sxan
X38	6-Thioxanthine	6Sxan

TABLE 1

Base no.	Base	Metal	Stoichiometry	Method	Ref.
			Adenine		
A1	ade	Cd(II)	$[Cd(ade)_2Cl_2]$	ir, esr	41
				ir, cond, nmr, therm	203
				spec	31
				ir, therm	23
			$[Cd(ade)_2SO_4]$	ir, cond, nmr, therm	203
			$[Cd(\mu\text{-ade})(\mu\text{-}H_2O)(NO_3)_2]_2(NO_3)_2$	X-ray	45
		Co(II)	Co...ade	ks, titr, pk	9
				ks	10
				pot, ks	36
			$[Co(ade)Cl_2]$	mag, ir, esr, spec	41
			$Co...(ade)_2$	ks	10
			$(ade)_2[Co(ade)_2(OH_2)_4](SO_4)_2$	X-ray	241
			$[Co(ade)_2Cl_2]$	mag, ir, esr, spec	41
			$[Co(ade)_2(ClO_4)_2].3EtOH$	mag, spec	5
		Co(III)	$[Co(Bu_3P)(ade)(dmg)_2]$	nmr	96
			$[Co(en)_2(ade)Cl]Br$	X-ray	256
				X-ray	257
		Cr(III)	$[Cr_3(ade)_5Cl_4]$	mag, ir, spec	26
		Cu(II)	Cu...ade	ks, titr, pk	9
				pot, ks	10
				pot, ks	36
			$Cu...D_2O...ade$	ir	115
			Cu...nta...ade	ks	42
			$[Cu(ade)Cl_2]$	mag, ir, esr, spec	41
			$[Cu(ade)(H_2O)(glygly)]$	X-ray	242
			$Cu...(ade)_2$	pot, ks	10
			$[Cu(ade)_2Br_2]$	spec	31
			$[Cu(ade)_2Br_2]Br_2$	X-ray	191
				X-ray	199
			$[Cu(ade)_2Cl_2]$	spec	31
				mag, ir, esr, spec	41
				spec	243
			$[Cu(ade)_2(ClO_4)_2].EtOH$	mag, spec	5
				mag, spec	31
			$[Cu(ade)_2(NO_3)_2]$	spec	31
			$[Cu(ade)_4(H_2O)_2]$	X-ray	255
				esr	119
				X-ray	254
			$[Cu_2(ade)_4(H_2O)_2](ClO_4)_2$	esr	119
			$[Cu_2(ade)_4(H_2O)_2](SO_4)_2$	esr	119

TABLE 1 (CONTINUED)

Base no.	Base	Metal	Stoichiometry	Method	Ref.
			$[Cu_2(\mu\text{-ade})_4(H_2O)_2](ClO_4)_4$	X-ray	245
				mag	246
			$[Cu_2(ade)_4Cl_2]^{2+}$	X-ray	244
			$[Cu_2(ade)_4Cl_2]$	esr	119
			$[Cu_2(ade)_4(NH_3)_2]$	esr	119
			$[Cu_2(ade)_4(pip)_2]$	esr	119
			$[Cu_3(ade)_2Cl_8]$	X-ray	258
				X-ray	259
		Dy(III)	$[Dy(ade)_2Cl_2]$	mag, ir	22
		Au(III)	$[Au(ade)_2Cl_3]$	cond, uv, ir	34
		Fe(II)	$[Fe(ade)(ClO_4)]$.EtOH	mag, spec	5
		Fe(III)	$[Fe(ade)_2Cl_3]$	mag, ir, spec	25
		Hg(II)	$[Hg(ade)_2Cl_2]$	ir, esr	41
			$[Hg_2(ade)_3Cl_2]$	ir, therm	23
			[(MeHg)(ade)]	ir, X-ray	32
			$[(MeHg)_2(ade)]$	ir, X-ray	37
			$[(MeHg)_2(ade)]ClO_4$	ir	32
			$[(MeHg)_2(ade)]$.EtOH	ir	215
			$[(MeHg)_3(ade)]$	X-ray, ir	37
				X-ray	215
			$[(MeHg)_3(ade)](ClO_4)_3$	ir	32
			$[(MeHg)_4(ade)](NO_3)$	X-ray	215
			$[(MeHg)_5(ade)_2]$	X-ray	11
		Ir(III)	$[Ir_3(ade)_4Cl_9]$.3MeOH	ir	18
		Mn(II)	$[Mn(ade)_2(ClO_4)_2]$.2EtOH	mag, spec	5
		Ni(II)	Ni...ade	ks, titr, pk	9
				ks	10
				pot, ks	36
			$[Ni(ade)Cl_2]$	mag, ir, esr, spec	41
			$Ni...(ade)_2$	ks	10
			$[Ni(ade)_2Cl_2]$	mag, ir, esr, spec	41
				spec	31
			$[Ni(ade)_2(ClO_4)]$.EtOH	mag, spec	5
		Pd(II)	$[Pd(ade)Cl_2]$	ir	21
			$[Pd(ade)_2Cl_2]$	ir	43
			$[Pd(ade)_2(nBu3P)_2]$	X-ray, ir	265
			$[Pd_2(ade)_3Cl_2$	ir, therm	23
		Pt(II)	$[Pt(NH_3)_2Cl_2]$...ade	HPLC	176
			$[Pt(ade)(NH_3)_2](NO_3)_2$	ir	48
				ir	49
			$[Pt(NH_3)_2(ade)(gua)]^{2+}$	kin, uv, eph, HPLC	73
			$[Pt(dien)Cl]^+$...ade	kin	210
			$[Pt(en)Cl_2]$...ade	chr, ra, uv, kin	40
				HPLC, kin	176

TABLE 1 (CONTINUED)

Base no.	Base	Metal	Stoichiometry	Method	Ref.
			$[Pt(NH_3)_3(ade)]$	calc	214
			$[Pt(ade)Cl_2]$	uv, ir, nmr, atta	15
				uv, ir, nmr, therm, atta	264
			$[Pt(ade)I_2]$	uv, ir, nmr	264
			$[Pt(ade)(H_2O)Cl_2]$	nmr, ir	17
			$[Pt(ade)(gly)Cl_2]$	nmr, spec, ir	16
			$[Pt(ade)(ala)Cl_2]$	nmr, spec, ir	16
			$[Pt(ade)Cl_3]$	nmr, ir	17
			$[Pt(ade)_2Cl_4]$	ir	21
			$[Pt(ade)_2(py)_2I_2]$	spec, ir	8
			$[Pt(ade)_2(thf)_2I_3]$	spec	8
			$[Pt(ade)_4][PtI_4]$	spec	8
			$[Pt_2(ade)_2(NH_3)_4Cl]Cl_2$	ir	215
			$[Pt_2(ade)_2(DMSO)_3I_2]$	spec, ir	8
			$[Pt_2(ade)_2(DMSO)_3I_3]$	spec, ir	8
			$[Pt_2(ade)_2(dmf)_2I_3]$	spec, ir	8
			$[Pt_2(ade)_2(dmf)_3I_3]$	spec, ir	8
			$[Pt_2(ade)_3Cl_4(HCl)_3]$	ir	175
		Pt(IV)	$[Pt(ade)Cl_5]$	spec, nmr, ir	17
			$[Pt_2(ade)(\mu\text{-}Cl)Cl_6]$	spec, nmr, ir	17
			$[Pt(ade)_3Cl_3]Cl$	nmr, ir	17
		Rh(I)	$[Rh(PPh_3)_2(ade)(CO)]PF_6$	nmr	44
		Rh(III)	$[Rh(ade)Cl_3].EtOH$	uv, ir	34
			$[Rh(ade)_2Cl_3].MeOH$	ir	18
		Ru(II)	$[Ru(ade)Cl_2(DMSO)_3]$	ir, nmr	20
			$[Ru(ade)_2(DMSO)_4](BF_4)_2$	cond, ir, nmr	20
			$[Ru(ade)_2(DMSO)_4]Cl_2$	cond, ir, nmr	20
		Ru(III)	$[Ru(ade)_2Cl_3].MeOH$	ir	18
			$[Ru(ade)_2(MeOH)(ClO_4)_3].MeOH$	cond, spec, ir	204
			$[Ru(ade)_3(DMSO)_3]Cl_4$	ir	18
		Th(IV)	$[Th(ade)_2Cl_2]$	mag, ir	22
		U(IV)	$[U(ade)_2Cl_2]$	mag, ir	22
			$[UO_2(ade)_2(H_2O)_2]$	ir, therm	23
		V(III)	$[V(ade)_3Cl].EtOH$	mag, ir, spec	27
		V(IV)	$[VO(ade)Cl_2]$	ir, spec	24
		Zn(II)	Zn...ade	ks	10
			$[Zn(ade)Cl]$	ir, esr	41
				ir	39
			$[Zn(ade)Cl_2]$	ir	39
			$[Zn(ade)Br]$	ir	39
			$[Zn(ade)I]$	ir	39
			$[Zn(ade)S]$	ir	39
			$[Zn(ade)ClO_4].EtOH$	mag, spec	5
				ir	39
			$[Zn(ade)(BF_4)].EtOH$	ir	39
			$[Zn(ade)(NO_3)].EtOH$	ir	39

TABLE 1 (CONTINUED)

Base no.	Base	Metal	Stoichiometry	Method	Ref.
			[Zn(ade)(NCS)].EtOH	ir	39
			$[Zn(ade)(ac)_2]$	ir	39
			$[Zn(ade)Cl_3]$	X-ray	198
				X-ray	197
			$Zn...(ade)_2$	ks	10
			$[Zn(ade)_2Cl_2]$	ir, esr	41
				spec	31
			Adenine-N(1)-oxide		
A2	ade-lox	Co(II)	$[Co(adelox)_2(ClO_4)_2].2EtOH$	ir, uv, mag	29
		Cr(III)	$[Cr(ade1ox)_2Cl_2]$	ir, spec, mag	145
			$[Cr(ade1ox)_2(ClO_4)_3].2EtOH$	ir, uv, mag	29
		Cu(II)	$Na_2[Cu(ade1ox)_2]$	X-ray	2
				X-ray	3
			$[Cu_2(ade1ox)_3(ClO_4)_2]$	ir, uv, mag	29
			$Cu(II)...ade1ox...H_3O^+$	X-ray	196
				X-ray	195
		Dy(III)	$[Dy(ade1ox)_3Cl_3]_2$	ir, spec, mag	145
		Fe(III)	$[Fe(ade1ox)_2Cl_2]$	ir, spec, mag	145
			$[Fe(ade1ox)_2Cl_3]$	ir, spec, mag	145
			$[Fe(ade1ox)_3(ClO_4)_2]$	ir, uv, mag	29
		Hg(II)	$[Hg(ade1ox)Cl_2]$	X-ray	19
		Mn(II)	$[Mn(ade1ox)_2(ClO_4)_2].2EtOH$	ir, uv, mag	29
		Ni(II)	$[Ni(ade1ox)_2(ClO_4)].2EtOH$	ir, uv, mag	29
		Th(IV)	$[Th(ade1ox)_2Cl_3]$	ir, spec	145
		U(IV)	$[U(ade1ox)_2Cl_3]$	ir, spec, mag	145
		V(IV)	$[VO(ade1ox)Cl_2]$	ir, spec, mag	145
		Zn(II)	$[Zn(ade1ox)_2(ClO_4)_2].2EtOH$	ir, uv	29
			8-Azaadenine		
A3	8azade	Co(II)	Co...8azade	cond, ks	36
		Cu(II)	Cu...8azade	cond, ks	36
				X-ray	194
		Hg(II)	[(MeHg)(8azade)]	X-ray	6
			$[(MeHg)(8azade)]NO_3$	X-ray	6
		Ni(II)	Ni...8azade	cond, ks	36
			3-Benzyladenine		
A4	3bzade	Co(III)	$[Co(3bzade)(acac)_2(NO_2)]$	nmr	7
			$[Co(3bzade)(dmg)_2(et)]$	nmr	117
			$[Co(3bzade)(dmg)_2(ipr)]$	nmr	117
			$[Co(3bzade)(dmg)_2(me)]$	nmr	117
			$[Co(3bzade)(dmg)_2(CH_2Br)]$	nmr	117
			$[Co(3bzade)(dmg)_2(OCH_3)_2P(O)]$	nmr, X-ray	117

TABLE 1 (CONTINUED)

Base no.	Base	Metal	Stoichiometry	Method	Ref.
			3-((Ethoxycarbonyl)methyl)adenine		
A5	ecma	Co(III)	$[Co(ecma)(dmg)_2(ipr)]$	nmr	117
			$[Co(ecma)(dmg)_2(me)]$	nmr	117
			$[Co(ecma)(dmg)_2(CH_2Br)]$	nmr	117
			$[Co(ecma)(dmg)_2P(O)(OCH_3)_2]$	nmr	117
			3-Methyladenine		
A6	3me-ade	Co(III)	$[Co(3meade)(acac)_2]$	nmr	7
			[Co(3meade)(dmg)(Me)]	nmr	117
			$[Co(3meade)(dmg)P(O)(OCH_3)_2]$	nmr	117
		Pt(II)	$[Pt(NH_3)_2(3meade)_2](NO_3)_2$	X-ray, nmr	12
			9-Methyladenine		
A7	9me-ade	Cd(II)	$[Cd(9meade)Br_2]$	spec	31
			$[Cd(9meade)Cl_2]$	spec	31
		Co(II)	$[Co(9meade)Br_2]$	spec	31
			$[Co(9meade)Cl_2]$	spec	31
				X-ray	251
			$[Co(9meade)_2(NO_3)_2]$	spec	31
		Co(III)	$[Co(9meade)(dmg)_2(ipr)]$	nmr	117
			$[Co(9meade)(dmg)_2(me)]$	nmr	117
			$(Co(9meade)(dmg)_2P(O)(OCH_3)_2]$	nmr	117
		Cu(II)	Cu...9meade	pk	123
			$[Cu(9meade)Br_2]$	spec	31
			$[Cu(9meade)Cl_2]$	spec	31
			$[Cu(9meade)(NO_3)_2]$	spec	31
			$[Cu(9meade)(glygly)(H_2O)]$	X-ray	235
			$[Cu(9meade)(NMeN'SEN)(H_2O)]NO_3$	X-ray	248
			$[Cu(9meade)(OH_2)_4(SO_4)]_4$	X-ray	249
			$[Cu(9meade)_2(OH_2)_4]Cl_2$		250
		Hg(II)	$[(MeHg)(9meade)]NO_3$	X-ray	38
			$[(MeHg)_2(9meade)]$	X-ray	33
		Mo(II)	$[(C_5H_5)_2Mo(9meade)]PF_6$	nmr, X-ray	72
		Ni(II)	Ni...9meade	ks, pk	123
				ks, pk	124
			$[Ni(9meade)Br_2]$	spec	31
			$[Ni(9meade)Cl_2]$	spec	31
			$[Ni(9meade)(NO_3)_2]$	spec	31
		Pt(II)	$[Pt(9meade)Cl_3]$	X-ray	33
				X-ray	28
				ram	30
				therm	13
				nmr, ir	219
			$[Pt(NH_3)_3(9meade)]^{3+}$	X-ray, nmr	35

TABLE 1 (CONTINUED)

Base no.	Base	Metal	Stoichiometry	Method	Ref.
			$[Pt(NH_3)_2(9meade)]^+...CN^-$	kin	135
			$[PtCl_2(DIPSO)_2(9meade)$		238
			$[Pt(dien)(9meade)]^{2+}$	nmr, uv	218
			$[(Pt(dien))_2(\mu\text{-}9meade)]^{4+}$	nmr, uv	218
			$[(PtCl_2)_2(DMSO)_2(\mu\text{-}9meade)]$	nmr	220
				X-ray	238
			$[Pt(NH_3)_2(9meade)(lmecyt)]ClO_4$	nmr, HPLC	4
			$[Pt(NH_3)_2(9meade)(1mecty)]^{2+}...CN^-$	kin	174
			$[Pt(NH_3)_2(9meade)(1mecyt)](ClO_4)_2$	nmr, X-ray	1
				nmr, HPLC	4
			$[Pt(NH_3)_2(9meade)(1mecyt)](ClO_4)_3$	nmr	1
			$[Pt_2(NH_3)_4(\mu\text{-}9meade)(1mecyt)_2]^{4+}$	nmr	1
			$[Pt(NH_3)_2(9meade)(1methy)]ClO_4$	nmr	1
			$[Pt(NH_3)_2(9meade)(1methy)]^{2+}$	nmr	137
			$[Pt_2(NH_3)_4(\mu\text{-}9meade)(1methy)_2]^{3+}$	nmr	1
			$[Pt(NH_3)_2(9meade)]^+$	pk	134
			$[Pt(NH_3)_2(9meade)_2]^{2+}$	nmr, uv	218
			$[Pt(NH_3)_2(9meade)_2]^+...CN^-$	kin	135
		Zn(II)	$[Zn(9meade)Br_2]$	spec	31
			$[Zn(9meade)Cl_2]$	spec	31
				X-ray	252
			$(9meade)[Zn(9meade)Cl_3]$	spec	253
			1,9-Dimethyladenine		
A8	19dme-ade	Co(II)	$[Co(19dmeade)(acac)(NO_2)]$ $[Co(acac)_2(NO_2)_2]$	X-ray	171
			2,9-Dimethyladenine		
A9	29dme-ade	Cu(II)	Cu...(2,9dmeade)	pk	123
		Ni(II)	Ni...(2,9dmeade)	pk	123
				pk	124
			8,9-Dimethyladenine		
A10	89dme-ade	Cu(II)	Cu...(89dmeade)	pk	123
		Ni(II)	Ni...(89dmeade)	pk	123
				pk	124
			3-(γ,γ-Dimethylallyl)adenine (tricanthine)		
A11	tct	Co(III)	$[Co(tct)(acac)_2(NO_2)]$	X-ray, nmr	7
			Caffeine		
X30	caf	Ag(I)	$[Ag(caf)](NO_3)$	therm, nmr	57

TABLE 1 (CONTINUED)

Base no.	Base	Metal	Stoichiometry	Method	Ref.
		Au(II)	$[Au(caf)Cl_4]$	ir, nmr, mag, therm	82
		Cu(I)	$[Cu(caf)_2(ClO_4)_2]$	cond, ir, spec, mag	77
		Hg(I)	$[Hg_2(caf)_2(NO_3)_2]$	X-ray	53
				ir, nmr, therm	68
		Pd(II)	$[Pd(caf)_2Cl_2]$	ir	67
				ir, nmr, therm	81
				ir, nmr	88
			$[Pd(caf)(ino)Cl]$	cond, nmr, ir	122
			$[Pd(caf)(ino)Cl_2]$	cond, nmr, ir	122
			$[Pd(caf)(guo)Cl]$	cond, nmr, ir	122
			$[Pd(caf)(guo)Cl_2]$	cond, nmr, ir	122
			$[Pd(caf)(ado)Cl_2]$	cond, nmr, ir	122
			$K[Pd(caf)Cl_3]$	cond, nmr, ir	122
		Pt(II)	$[Pt(caf)Cl_2]$	cond, nmr, ir	122
			$[Pt(caf)_2Cl_2]$	X-ray, ir	67
			$[Pt(caf)(cyd)Cl_2]$	ir	67
			$[Pt(caf)(ino)Cl]$	cond, nmr, ir	122
			$[Pt(caf)(ino)Cl_2]$	cond, nmr, ir	122
			$[Pt(caf)(guo)Cl]$	cond, nmr, ir	122
			$[Pt(caf)(guo)Cl_2]$	cond, nmr, ir	122
			$[Pt(caf)(ado)Cl_2]$	cond, nmr, ir	122
			$K[Pt(caf)Cl_3]$	cond, nmr, ir	122
				nmr, ir	67
		Ru(III)	$[Ru(caf)(NH_3)_3Cl_2]Cl$	X-ray	95
				uv, ech, pk	166
			$[Ru(caf)(NH_3)_4Cl]Cl_2$	uv, ech, pk	166
			Guanine		
G1	gua	Co(II)	$[Co(gua)Cl_2]$	ir, spec, mag	162
			$[Co(gua)_2Cl]$	ir, spec	159
			$[Co(gua)_2(EtOH)(H_2O)_2]ClO_4$	spec, ir, mag, cond	155
		Cr(III)	$[Cr(gua)_2Cl_3]$	ir, spec, mag	160
			$[Cr(gua)_2(ClO_4)_2(EtOH)_2]ClO_4$	spec, ir, mag, cond	155
		Cu(II)	$[Cu(gua)Cl_2]$	ir, spec, mag	162
			$[Cu(gua)_2]$	ir, spec, mag	86
			$[Cu(gua)_2Cl]$	ir, spec	159
			$[Cu(gua)_2(ClO_4)_2].2EtOH$	ir, spec, mag	86
			$[Cu(gua)Cl_3]_2$	X-ray	190
				X-ray	189
				X-ray	188
		Dy(III)	$[Dy(gua)_3Cl_2]$	ir, spec, mag	156
		Fe(II)	$[Fe(gua)(H_2O)_4]ClO_4$	spec, ir, mag, cond	155

TABLE 1 (CONTINUED)

Base no.	Base	Metal	Stoichiometry	Method	Ref.
			$[Fe(gua)Cl(ROH)_2]$	ir, spec	50
		Fe(III)	$[Fe(gua)_2Cl_3]$	ir, spec, mag	160
			$[Fe(gua)_2(ClO_4)(EtOH)_2]ClO_4$	spec, ir, mag, cond	155
		Hg(II)	$[(MeHg)(gua)]$	nmr	116
			$[(MeHg)_2(gua)]$	nmr	116
		Mn(II)	$[Mn(gua)Cl(ROH)_2]$	ir, spec	50
			$[Mn(gua)_2(EtOH)_3]$	ir, mag, cond	155
		Ni(II)	$[Ni(gua)Cl(ROH)_2]$	ir, spec	50
			$[Ni(gua)_2(EtOH)_3](ClO_4)_2$	spec, ir, mag, cond	155
		Pt(II)	$[Pt(NH_3)(gua)(tu)_2]^{2+}$	kin, eph, chr	73
			$[Pt(NH_3)_2(gua)(H_2O)]^{2+}$	kin, uv, eph, HPLC	73
			$[Pt(NH_3)_2(gua)Cl]Cl$	uv	46
				uv	47
			$[Pt(NH_3)_2(gua)(ade)]^{2+}$	kin, uv, eph, HPLC	73
			$[Pt(NH_3)_2(gua)(cyt)]^{2+}$	uv	47
			$[Pt(NH_3)_2Cl_2]$...gua	HPLC	237
			$[Pt(en)Cl_2]$...gua	chr, ra, uv, kin	40
			[Pt(dien)Cl]...gua	kin	210
			$[Pt(gua)Cl_2]$	atta, uv, ir, nmr	15
				atta, uv, ir, therm, nmr	264
			$[Pt(gua)I_2]$	uv, ir, nmr	264
			$[Pt(NH_3)_3(gua)]$	calc	214
			$[Pt(gua)Cl_3]MeOH$	spec, nmr, ir	17
			$[Pt(gua)(gly)Cl_2]$	nmr, spec, ir	16
			$[Pt(gua)(ala)Cl_2]$	nmr, spec, ir	16
			$[Pt(gua)_2Cl_2]$	spec, nmr, ir	17
			$[Pt(NH_3)_2(gua)_2]^{2+}$	chr	72
				uv, eph, kin, HPLC	46
				uv	47
			$Na_{10}[Pt_2(gua)_2(\mu\text{-}PO_4)_4]$	X-ray	211
				X-ray	212
		Pt(IV)	$[Pt(gua)Cl_5]$	spec, nmr, ir	17
			$[Pt(gua)_3Cl_3]Cl$	spec, nmr, ir	17
		Ru(III)	$[Ru(NH_3)_5(gua)]^{2+}$	HPLC	127
			$[Ru(gua)(H_2O)_5](ClO_4)_3$	nmr, ir, spec, cond	204
		Th(III)	$[Th(guo)_2Cl_2]$	ir, spec, mag	156
		U(IV)	$[U(gua)_2Cl_2]$	ir, spec, mag	156
		V(III)	$[V(gua)Cl_2(EtOH)_2]$	ir	157
		V(IV)	$[VO(gua)Cl_2]$	ir, spec, mag	160

TABLE 1 (CONTINUED)

Base no.	Base	Metal	Stoichiometry	Method	Ref.
		Zn(II)	$[Zn(gua)Cl_2]$	ir, spec	162
			$[Zn(gua)Cl_3]$	X-ray	197
			$[Zn(gua)_2Cl]$	ir, spec	159
			$[Zn(gua)_2(EtOH)_3](ClO_4)_2$	spec, ir, cond	155
			9-Ethylguanine		
G2	9etgua	Pt(II)	$[Pt(NH_3)_2Cl_2]$...9etgua	nmr, uv	167
			$[Pt(dien)Cl_2]$...9etgua	nmr, uv	167
			$[Pt(NH_3)(9etgua)(tu)_2]^{2+}$	nmr	174
			$[Pt(dien)(9etgua)]^{+}$	nmr, uv	167
			$[Pt(dien)(9etgua)]^{2+}$	nmr, uv	167
			$[(Pt(dien)(\mu\text{-}9etgua)]^{3+}$	nmr	167
			$[Pt(NH_3)_2(9etgua)D_2O](NO_3)_2$	nmr	169
			$[Pt(NH_3)_2(9etgua)OH]^{+}$	nmr	167
			$[Pt(NH_3)_2(9etgua)Cl]Cl$	nmr, ir	169
				nmr	167
				nmr	151
			$[Pt(NH_3)_2(9etgua)Cl]Cl...CN^-$	kin, ir, nmr	174
			$[Pt(NH_3)_2(9etgua)Cl]NO_3$	nmr	169
			$[Pt(NH_3)_2(9etgua)(H_2O)]^{2+}$	pk, titr, nmr	151
			$[Pt(NH_3)_2(9etgua)(1mecyt)]ClO_4$	nmr	153
				X-ray, ir, ram	152
			$[Pt(NH_3)_2(9etgua)(1mecyt)](ClO_4)_2$	nmr	154
				nmr	169
				X-ray, ir, ram	152
				nmr	153
				X-ray	225
				X-ray, nmr	239
			$[Pt(NH_3)_2(9etgua)(1mecyt)](ClO_4)_2...CN^-$	kin	174
			$[Pt(NH_3)_2(9etgua)(1mecyt)](ClO_4)_3$	X-ray, ir, ram	152
				nmr	153
			$[Pt(NH_3)_2(9etgua)(1methy)]ClO_4$	nmr	4
			$[Pt(NH_3)_2(9etgua)(1meura)]ClO_4$	nmr	169
			$[Pt(NH_3)_2(9etgua)(1meura)]ClO_4...CN$	kin	174
			$[Pt(NH_3)_3(9etgua)]^{2+}$	nmr	150
			$[Pt(NH_3)_3(9etgua)]Cl_2$	nmr	169
			$[Pt(NH_3)_3(9etgua)]^{2+}...CN^-$	kin	135
				kin	174
			$[Pt(NH_3)_2(9etgua)_2]$...9etgua		137
			$[Pt(NH_3)_2(9etgua)_2]Cl_2$	nmr, X-ray, ir	168
				nmr	169
				ram	161

TABLE 1 (CONTINUED)

Base no.	Base	Metal	Stoichiometry	Method	Ref.
			$[Pt(NH_3)_2(9etgua)_2]Cl_2$		
			...CN^-	kin	174
			...tu	kin	174
			$[Pt(NH_3)_2(9etgua)_2]Cl_{0.5}(HCO_3)_{1.5}$	nmr, X-ray, ir	168
			$[Pt(NH_3)_2(9etgua)_2](ClO_4)_2$	nmr	154
				nmr	153
				X-ray, ram	161
				nmr	167
			$[Pt(NH_3)_2(9etgua)_2](NO_3)_2$	ram	161
				nmr	174
			$[Pt(NH_3)_2(9etgua)_2]SO_4$	ram	161
			$[Pt(NH_3)(9etgua)_2(1mecyt)](ClO_4)_2$	nmr	226
			$[Pt(NH3)(9etgua)_2(1mecyt)](ClO_4)_2...CN$	kin	174
			$[Pt(pa)_2(9etgua)_2]^{2+}$		137
			$[Pt(NH_3)_2(9etgua)_2][Pt(CN)_4]$	X-ray, ram	161
			$[Pt(NH_3)_2(9etgua)_2][Pt(CN)_4]$	nmr	174
			$[Pt(NH_3)_2(9etgua)_2(\mu\text{-}9etgua)]^{3+}$	nmr	167
			$[Pt(NH_3)_2(9etgua)_3]$		137
			$[Pt_2(NH_3)_2(9etgua)(CN)_4]$	ir	174
			$[Pt_2(NH_3)_6(\mu\text{-}9etgua)]^{3+}$	nmr	150
			$[(Pt(NH_3)_3)_3(9etgua)]^{5+}$	nmr, X-ray	150
		Zn(II)	$[Zn(9etgua)Cl_2(OH_2)]$	X-ray	192
			1-Methylguanine		
G3	1megua	Ru(III)	$[Ru(NH_3)_5(1megua)]Cl_3$	uv, ech, pk	164
			9-Methylguanine		
G4	9megua	Cu(II)	$[Cu(9megua)_2(OH_2)_3]$	X-ray	193
		Pt(II)	$[Pt(9megua)(dach)]^{2+}$	CD, nmr	222
			$K_2[Pt(9megua)Cl_3]$	X-ray	223
			$[Pt(9megua)_2(dach)]^{2+}$	X-ray	224
			$[Pt(9megua)_2(1,3dach)]^{2+}$	X-ray	224
			$[Pt(9megua)_4]^{2+}$	X-ray	261
			***N*(2),*N*(2)-Dimethyl-9-methylguanine**		
G5	tmegua	Pt(II)	$[Pt(NH_3)_2(tmegua)Cl]PF_6$	X-ray	163
			***N*(2),*N*(2)-Dimethyl-9-propylguanine**		
G6	dmepr-gua	Pt(II)	$[Pt(NH_3)_2(dmeprgua)(H_2O)]^{2+}$	titr, nmr	151
			$[Pt(NH_3)_2(dmeprgua)Cl]^+$	nmr	151

TABLE 1 (CONTINUED)

Base no.	Base	Metal	Stoichiometry	Method	Ref.
			1,9-Dimethylguanine		
G7	19dme-gua	Ru(III)	$[Ru(NH_3)_5(19dmegua)]Cl_3$	pk, uv, ech	164
			7,9-Dimethylguanine		
G8	79dme-gua	Pt(II)	$[Pt(dien)(79dmegua)](PF_6)_2$	X-ray	118
			6-Selenoguanine		
G9	6Segua	Pt(II)	$[Pt(6Segua)(NH_3)_2]$	atta, uv, ir, CD	65
			6-Thioguanine		
G10	6Sgua	Hg(II)	[(MeHg)(6Sgua)]	nmr	116
			$[(MeHg)(6Sgua)_2]$	nmr	116
		Pt(II)	$[Pt(6Sgua)_2]$	cond, uv, ir	61
			$[Pt(6Sgua)_2Cl_2]$	cond, uv, ir	61
			$[Pt(NH_3)_2(6Sgua)]$	atta, uv, ir, OR	65
		Pt(IV)	$[Pt(6Sgua)_2Cl_2]$	uv, ir	61
			$[Pt(6Sgua)_2Cl_4]$	cond, uv, ir	61
			8-Thioguanine		
G11	8Sgua	Hg(II)	[(MeHg)(8Sgua)]	nmr	116
			$[(MeHg)_2(8Sgua)]$	nmr	116
			$[(MeHg)_3(8Sgua)]$	nmr	116
			Hypoxanthine		
H1	hyp	Co(II)	Co...hyp	ks	10
			$Co...(hyp)_2$	ks	10
			$Co(hyp)(SO_4).5H_2O$	X-ray, ir, spec, therm	205
		Co(III)	$[(Bu_3P)Co(hyp)(dmg)_2]$	nmr	96
		Cu(II)	Cu...hyp	ks	10
			Cu...nta...hyp	ks	42
			$[Cu(hyp)(SO_4)(H_2O)]$	X-ray, ir, spec, therm	205
			$[Cu(hyp)(ClO_4)(EtOH)_4]$	ir, spec, mag	86
			$Cu...(hyp)_2$	ks	10
			$[Cu_2(hyp)_4Cl_2]Cl_2$	esr	119
				X-ray	187

TABLE 1 (CONTINUED)

Base no.	Base	Metal	Stoichiometry	Method	Ref.
			$[Cu_2(hyp)_4(OH_2)_2]^{4+}$	X-ray	246
		Hg(II)	[(MeHg)(hyp)]	nmr	116
			$[(MeHg)(hyp)_2]$	nmr	116
			$[(MeHg)_3(hyp)]$	nmr	116
		Ni(II)	Ni...hyp	pot, ks	10
			$Ni(hyp)(SO_4).5H_2O$	ir, X-ray, spec, therm	205
			$Ni...(hyp)_2$	pot, ks	10
		Pd(II)	$[Pd(dien)Cl]^+$...hyp	kin, uv	210
			$[Pd(hyp)_2Cl]$	ir, spec	177
		Pt(II)	[Pt(en)(hyp)Cl]Cl	ir, nmr, spec, cond, pk	206
			$[Pt(en)(hyp)_2]Cl_2$	ir, nmr, spec, cond, pk	206
			$[Pt(dien)Cl]^+$...hyp	kin, uv	[210]
			$[Pt_2(\mu\text{-hyp})(ala)_2Cl_4]$	nmr, ir	16
			$[Pt_2(\mu\text{-hyp})(gly)_2Cl_4]$	nmr, ir	16
			$[Pt_3(hyp)_4Cl_6(HCl)]$	ir	175
		Pt(IV)	$[Pt(hyp)_2Cl_3]$	ir, spec	177
		Ru(II)	$[Ru(NH_3)_5(hyp)]$	spec, ech	173
		Ru(III)	$[Ru(NH_3)_5(hyp)]^{2+}$	HPLC	127
			$[Ru(NH_3)_5(hyp)]Cl_3$	X-ray, spec, ech, pk	173
				X-ray, pk	14
			$[Ru(hyp)_4(H_2O)_2](ClO_4)_3$	nmr, ir, cond, spec	204
		Zn(II)	Zn...hyp	ks	10
			$Zn...(hyp)_2$	ks	10
			8-Azahypoxanthine		
H2	8azhyp	Cd(II)	$[Cd(8azhyp)_2(OH_2)_4]$	X-ray	180
			1-Methylhypoxanthine		
H3	1me-hyp	Ru(II)	$[Ru(NH_3)_5(1mehyp)]^{n+}$	spec, ech	173
		Ru(III)	$[Ru(NH_3)_5(1mehyp)]Cl_3$	spec, ech, pk	173
			7-Methylhypoxanthine		
H4	7me-hyp	Ru(II)	$[Ru(NH_3)_5(7mehyp)]^{n+}$	spec, ech	173
		Ru(III)	$[Ru(NH_3)_5(7mehyp)]^{3+}$	HPLC	127
				X-ray, pk	14
			$[Ru(NH_3)_5(7mehyp)]^{2+}$	HPLC	127
			$[Ru(NH_3)_5(7mehyp)]Cl_3$	X-ray, spec, ech, pk	173

TABLE 1 (CONTINUED)

Base no.	Base	Metal	Stoichiometry	Method	Ref.
			7-Methyl-9-propylhypoxanthine		
H5	7me-9prhyp	Pt(II)	$[Pt(7me9prhyp)(DMSO)Cl_2]$	nmr	229
			$[Pt(7me9prhyp)_2(DMSO)Cl]^+$	nmr	229
			9-Methylhypoxanthine		
H6	9me-hyp	Cu(II)	$[Cu(9mehyp)_2(OH_2)_2Cl_2]$	X-ray	186
				X-ray	185
		Hg(II)	$[Hg(9mehyp)Cl_2]$	X-ray	93
		Pt(II)	$[Pt(NH_3)_2Cl_2]$...9mehyp	nmr, chr	99
			$[Pt(NH_3)_2(9mehyp)]^+$	nmr, chr	99
			$[Pt(dien)(9mehyp)]^+$	nmr	99
			$[Pt(dien)(9mehyp)]^{2+}$	nmr, chr	99
			[Pt(dmdap)(9mehyp)]	nmr, chr	207
			$[Pt(NH_3)_2(9mehyp)_2]$	chr, nmr	99
			$[Pt(NH_3)_2(9mehyp)_2]^+$	chr, nmr	99
			$[Pt(NH_3)_2(9mehyp)_2]^{2+}$	chr, nmr	99
				chr, nmr	207
			$[Pt(en)(9mehyp)_2]^{2+}$	nmr	207
			$[Pt(dmdap)(9mehyp)_2]^{2+}$	nmr	85
			$[Pt(tmtn)(9mehyp)_2]^{2+}$	nmr	85
			$[Pt(tmdap)(9mehyp)_2]^{2+}$	nmr	85
			$[Pt(dmtn)(9mehyp)_2]^{2+}$	nmr	85
			$[Pt(pic)_2(9mehyp)_2]^{2+}$	nmr	85
			$[Pt(bipy)(9mehyp)_2]^{2+}$	nmr	85
			$[Pt(bpe)(9mehyp)_2]^{2+}$	nmr	85
			$[Pt(bpe)(9mehyp)_2](NO_3)_2$	X-ray, nmr	112
			$[Pt(NH_3)_2Cl_2]_2$...9mehyp	nmr, chr	99
			$[(Pt(dien))_2(\mu\text{-}9mehyp)]^{3+}$	nmr, chr	99
			$[(Pt(NH_3)_2)_2(9mehyp)_2\mu\text{-}9mehyp)]$	nmr	99
		Pt(IV)	$[Pt(en)(9mehyp)Cl_3]$	nmr	207
			$[Pt(dmdap)(9mehyp)Cl_3]$	nmr	207
			$[Pt(en)(9mehyp)_2]^{2+}$	nmr	207
			$[Pt(dmdap)(9mehyp)_2]^{2+}$	nmr	227, 207
			7,9-Dimethylhypoxanthine		
H7	dme-hyp	Cu(II)	[Cu(dmehyp)(glygly)]	X-ray	51
		Pt(II)	$[Pt(dien)(dmehyp)]^{2+}$	nmr	229
			$[Pt(dien)(dmehyp)](PF_6)_2$	X-ray	71
				X-ray	118
			$[Pt(NH_3)_2(dmehyp)_2](NO_3)_2$	X-ray, nmr	158
			$[Pt(en)(dmehyp)_2]^{2+}$	nmr	229
			$[Pt(en)(dmehyp)_2](PF_6)$	X-ray	71
				X-ray	114

TABLE 1 (CONTINUED)

Base no.	Base	Metal	Stoichiometry	Method	Ref.
			1,4-Bis(hypoxanth-9-yl)butane		
H8	hypbu	Pt(II)	$[Pt(en)(\mu\text{-hypbu})]_2(PF_6)_4$	nmr, X-ray	100
			$[Pt(NH_3)_2Cl)_2(en)(\mu\text{-hypbu})]Cl_2$	nmr	100
			1,3-Bis(hypoxanth-9-yl)-2-propanol		
H9	hyppo	Pt(II)	[Pt(en)(hyppo)]Cl	nmr	100
			1,3-Bis(hypoxanth-9-yl)-2-propane		
H10	hyppr	Pt(II)	[Pt(en)(hyppr)]Cl	nmr	100
			Purine		
P1	pur	Cr(III)	$[Cr(pur)_5Cl]Cl_3$	ir, mag, spec	26
		Co(II)	$[Co(pur)Cl_2]$	ir, spec, mag	90
			$[Co(pur)_2(ClO_4)_2]$	ir, spec, mag, cond	120
		Cu(II)	$[Cu(pur)Cl_2]$	ir, spec, mag	90
			$[Cu(H_2O)_4(pur)]SO_4$	X-ray	129
			$[Cu_2Cl_6(pur)]$	X-ray	128
			$[Cu(pur)_2(ClO_4)_2]$	mag, ir, spec, cond	120
		Dy(III)	$[Dy(pur)_2Cl_2]$	mag, ir	22
		Fe(III)	$[Fe(pur)_2Cl_3]$	mag, ir, spec	25
			$[Fe(pur)_2(ClO_4)_2]$	mag, ir, spec, cond	120
		Mn(II)	$[Mn(pur)Cl_2]$	ir, spec, mag	90
			$[Mn(pur)_2(ClO_4)_2]$	ir, spec, mag, cond	120
		Ni(II)	$[Ni(pur)Cl_2]$	ir, spec, mag	90
			$[Ni(pur)_2(ClO_4)_2]$	ir, spec, mag, cond	120
		Pt(II)	$[Pt_2(pur)_3Cl_3(OH)]$	ir	175
		Ru(I)	$[Rh(PPh_3)_2(pur)(CO)]PF_6$	nmr	44
		Ti(IV)	$[(C_5H_5)_2Ti(pur)Cl]$	X-ray	125
		Th(IV)	$[Th(pur)_2Cl_2]$	mag, ir	22
		U(IV)	$[U(pur)_2Cl_2]$	mag, ir	22
		V(III)	$[V(pur)_3Cl].EtOH$	mag, ir, spec	27
		V(IV)	$[VO(pur)Cl_2]$	spec, ir	24
		Zn(II)	$[Zn(pur)Cl_2]$	ir, spec	90
			$[Zn(pur)Cl_3]$	X-ray	121
			$[Zn(pur)_2(ClO_4)_2]$	mag, ir, spec, cond	120

TABLE 1 (CONTINUED)

Base no.	Base	Metal	Stoichiometry	Method	Ref.
			Purine-*N*(1)-oxide		
P2	pur1ox	Cr(III)	$[Cr(pur1ox)_2(ClO_4)_3].2EtOH$	ir, mag, spec	94
		Co(II)	$[Co(pur1ox)_2(ClO_4)_2].2EtOH$	ir, mag, spec	94
		Cu(II)	$[Cu(pur1ox)_2(ClO_4)_2].EtOH$	ir, mag, spec	94
		Fe(II)	$[Fe(pur1ox)_2(ClO_4)_2].EtOH$	ir, mag, spec	94
		Fe(III)	$[Fe(pur1ox)_2(ClO_4)_3]$	ir, mag, spec	94
		Mn(II)	$[Mn(pur1ox)_2(ClO_4)_3].EtOH$	ir, mag, spec	94
		Ni(II)	$[Ni(pur1ox)_2(ClO_4)_2].EtOH$	ir, mag, spec	94
		Zn(II)	$[Zn(pur1ox)_2(ClO_4)_2].EtOH$	ir, spec	94
			2-Amino-9-methylpurine		
P3	2A9-mepur	Cu(II)	Cu...2A9mepur	pk	123
		Ni(II)	Ni...2A9mepur	pk	123
			8-Amino-9-methylpurine		
P4	8A9-mepur	Cu(II)	Cu...8A9mepur	pk	123
		Ni(II)	Ni...8A9mepur	pk	123
			6-Aminobenzylpurine		
P5	6Abz-pur	Cu(II)	$[Cu(6Abzpur)Cl_3].MeOH$	cond, ir, mag, spec, esr	178
			$[Cu_2(6Abzpur)_4(H_2O)_4].(H_2O)_4$	therm, esr, cond, ir, mag, spec	178
			$[Cu_2(6Abzpur)_4](ClO_4)_4$	therm, esr, cond, ir, mag, spec	178
			6-Aminoethylaminepurine		
P6	6Aea-pur	Pt(II)	$[Pt(6Aeapur)Cl_2]$	atta, nmr	106
			2,6-Diaminopurine		
P7	26dA-pur	Cu(II)	Cu...(26dApur)	ir	115
		Pt(II)	$[Pt(26dApur)Cl_3]$	cond, ir, uv	61
		Pt(IV)	$[Pt(26dApur)Cl_5]$	cond, ir, uv	61
			$[Pt_2(26dApur)_2(OH)_2Cl_6]$	ir, uv	61
			2-Chloro-9-methylpurine		
P8	2Cl9-mepur	Cu(II)	Cu...(2Cl9mepur)	pk	123
		Ni(II)	Ni...(2Cl9mepur)	pk	123

TABLE 1 (CONTINUED)

Base no.	Base	Metal	Stoichiometry	Method	Ref.
			8-Chloro-9-methylpurine		
P9	8Cl9-mepur	Cu(II)	Cu...(8Cl9mepur)	pk	123
		Ni(II)	Ni...(8Cl9mepur)	pk	123
			6-Hydroxyethylaminepurine		
P10	6HO-eapur	Pt(II)	[Pt(6HOeapur)Cl_2]	atta, nmr	106
			6-Hydroxyethylmethylaminepurine		
P11	6HO-emapur	Pt(II)	[Pt(6HOemapur)Cl_2]	nmr	106
			2-Methoxy-9-methylpurine		
P12	2meo9-mepur	Cu(II)	Cu...(2meo9mepur)	pk	123
		Ni(II)	Ni...(2meo9mepur)	pk	123
				pk	124
			8-Methoxy-9-methylpurine		
P13	8meo9-mepur	Cu(II)	Cu...(8meo9mepur)	pk	123
		Ni(II)	Ni...(8meo9mepur)	pk	123
				pk	124
			9-Methylpurine		
P14	9mepur	Co(II)	Co...(9mepur)	ks, pk	79
				uv, spec, ks	78
		Cu(II)	Cu...(9mepur)	ks, pk	79
				uv, spec, ks	78
			Cu...$(9mepur)_2$	uv, spec, ks	78
		Mn(II)	Mn...(9mepur)	ks, pk	79
		Ni(II)	Ni...(9mepur)	ks, pk	79
				uv, spec, ks	78
				ks, pk	124
			Ni...$(9mepur)_2$	uv, spec, ks	78
		Zn(II)	Zn...(9mepur)	ks, pk, nmr	79
				ks	78
			2,9-Dimethylpurine		
P15	29dme-pur	Co(II)	Co...(29dmepur)	ks, pk	79
		Cu(II)	Cu...(29dmepur)	ks, pk	79
				uv, spec, ks	78

TABLE 1 (CONTINUED)

Base no.	Base	Metal	Stoichiometry	Method	Ref.
			Cu...$(29dmepur)_2$	uv, spec, ks	78
		Mn(II)	Mn...(29dmepur)	ks, pk	79
		Ni(II)	Ni...(29dmepur)	ks, pk	79
				uv, spec, ks	78
				ks, pk	124
		Zn(II)	Zn...(29dmepur)	ks, pk, nmr	79
			6,9-Dimethylpurine		
P16	69dme-pur	Co(II)	Co...(69dmepur)	ks, pk	79
		Cu(II)	Cu...(69dmepur)	ks, pk	79
		Mn(II)	Mn...(69dmepur)	ks, pk	79
		Ni(II)	Ni...(69dmepur)	ks, pk	79
		Zn(II)	Zn...(69dmepur)	ks, pk	79
			8,9-Dimethylpurine		
P17	89dme-pur	Co(II)	Co...(89dmepur)	ks, pk	79
		Cu(II)	Cu...(89dmepur)	uv, spec, ks	78
				ks, pk	79
			Cu...$(89dmepur)_2$	uv, ks, spec	78
		Mn(II)	Mn...(89dmepur)	ks, pk	79
		Ni(II)	Ni...(89dmepur)	uv, spec, ks	78
				ks, pk	79
				ks, pk	124
		Zn(II)	Zn...(89dmepur)	ks, pk, nmr	79
			6-Thiopurine		
P18	6Spur	Cd(II)	$Cd(6Spur)Cl_2$	X-ray	231
			$Cd(6Spur)_2$	X-ray	146
			$Cd(6Spur)_4Cl_2$	X-ray	146
		Co(II)	[Co(6Spur)(ac)]	ir, therm, mag	102
		Cu(I)	$[Cu(6Spur)Cl_2]_2$	X-ray	146
				X-ray	233
				X-ray	234
				X-ray	179
		Fe(II)	$[Fe(6Spur)(OH)SO_4]$	ir, therm, mag, MB	102
		Hg(II)	$[Hg(6Spur)_2Cl_2]$	X-ray	230
			[(MeHg)(6Spur)]	nmr	116
			$[(MeHg)_2(6Spur)]$	nmr	116
		Pd(II)	$Na_2[Pd(6Spur)_2Cl_2]$	atta	236
		Pt(II)	$[Pt(NH_3)_2(6Spur)]$	atta, uv, ir, OR	65
		Pt(IV)	$Na_2[Pt(6Spur)Cl_4]$	atta	236
		Ru(II)	$[Ru(6Spur)_2(DMSO)_3]Cl_2$	ir, nmr, spec	104

TABLE 1 (CONTINUED)

Base no.	Base	Metal	Stoichiometry	Method	Ref.
			$[Ru(6Spur)_2(PPh_3)_2]Cl_2.2EtOH$	ir, nmr, X-ray	98
			$[Ru(6Spur)_2(PPh_3)_2]Cl$	ir, nmr	98
			$[Ru(6Spur)_4Cl_2]$	cond, ir, nmr, spec	104
		Ru(III)	$[Ru(6Spur)Cl_3]$	cond, mag, ir, nmr, spec	104
			$[Ru(6Spur)_2Cl_3]$	cond, mag, ir, nmr, spec	104
			9-Benzyl-6-thiopurine		
P19	9bz6S-pur	Pd(II)	$[Pd(9bz6Spur)_2(dma)]$	X-ray	216
				X-ray	217
			6-Butylthiopurine		
P20	6buS-pur	Pd(II)	$[Pd(6buSpur)_3Cl]Cl$	atta	236
			9-Methyl-6-thiopurine		
P21	9me6S-pur	Cu(II)	$[Cu(9me6Spur)Cl_2]_2$		232
		Ru(II)	$[Ru(9me6Spur)(PPh_3)_2]Cl_2$	nmr, ir	98
			1,9-Dimethyl-6-thiopurine		
P22	19dme-6Spur	Ru(II)	$[Ru(19dme6Spur)_2(PPh_3)_2]Cl_2$	nmr, ir	98
			9-Methyl-2-methylthiopurine (2,9-Dimethyl-2-thiopurine)		
P23	29dme-2Spur	Cu(II)	Cu...(29dme2Spur)	pk	123
		Ni(II)	Ni...(29dme2Spur)	pk	123
			9-Methyl-8-methylthiopurine (8,9-Dimethyl-8-thiopurine)		
P24	89dme-8Spur	Cu(II)	Cu...(89dme8Spur)	pk	123
		Ni(II)	Ni...(89dme8Spur)	pk	123
			Theobromine		
X20	thb	Ag(I)	[Ag(thb)]	ir, uv, nmr, pk	60
				ir, nmr, therm	68
			$[Ag(thb)]NO_3$	nmr, therm	57
		Au(III)	$[Au(thb)Cl_4]$	ir, nmr, mag, therm	82

TABLE 1 (CONTINUED)

Base no.	Base	Metal	Stoichiometry	Method	Ref.
		Cu(II)	$[Cu(thb)_2(ClO_4)_2]$	cond, ir, spec, mag	77
		Pd(II)	$[Pd(thb)_2Br_2]$	ir, therm	209
			$[Pd(thb)_2Br_4]$	ir, therm	209
			$[Pd(thb)_2Cl]$	ir, spec	177
			$[Pd(thb)_2Cl_2]$	ir, nmr, therm	81
				ir, nmr	88
			$[Pd(thb)_2Cl_4]$	ir, nmr, therm	70
				ir, nmr	88
		Ru(III)	$[Ru(NH_3)_4(thb)Cl]Cl_2$	uv, ech, pk	166
			Theophylline		
X10	thp	Ag(I)	[Ag(thp)]	ir, spec, ks	59
				ir, therm	83
		Au(III)	$[Au(thp)Cl_4]$	ir, nmr, mag, therm	82
		Cd(II)	$[Cd(thp)_2(OH_2)_4]$	ir, therm	63
				ir, nmr, therm	70
			$[Cd(thp)_2(NH_3)_2]$	ir	92
				ir, therm	83
				ir, spec	59
			$[Cd(thp)_2(NH_3)_2(H_2O)_2]$	ir	111
		Cr(III)	$[Cr(thp)_2(ClO_4)_2]ClO_4$	cond, mag, spec, ir	208
		Co(II)	$[Co(thp)_2]$	ir, spec, ks	59
				ir, therm	83
			$[Co(thp)_2(ClO_4)_2(H_2O)_2]$	ir	105
				ir, cond, mag, spec	208
		Co(III)	$[Co(en)_2(thp)Cl]Cl$	X-ray	55
			$[Co(en)_2(thp)Cl]^+$	X-ray	184
				X-ray	183
				X-ray	149
				X-ray	107
		Cu(II)	$[Cu(thp)(H_2O)_2Cl_2]$	X-ray	75
			$[Cu(thp)(H_2O)_2(NMeN'SEN)]$	X-ray	182
				X-ray	192
			[Cu(thp)(NMeN′SEN)]	X-ray	181
				X-ray	172
				X-ray, mag	170
			$[Cu(thp)(H_2O)_2Cl_2]$	X-ray	75
			$[Cu(thp)_2]$	ir, mag	111
			$[Cu(thp)_2(NH_3)_2]$	ir, spec, ks	59
				ir	92
				ir, therm	83
				ir, mag	111

TABLE 1 (CONTINUED)

Base no.	Base	Metal	Stoichiometry	Method	Ref.
			$[Cu(thp)_2(NH_3)_2(H_2O)_2]$	ir, mag	111
			$[Cu(thp)_2(NO_3)(H_2O)_2]^+$	X-ray	192
			$[Cu(thp)_2Br_2]$	ir	111
			$[Cu(thp)_2Cl_2]$	ir, mag	111
			$[Cu(thp)_2Cl_4]$	therm, mag	58
			$[Cu(thp)_2(ClO_4)_2]$	cond, ir, spec, mag	77
				ir	105
			$[Cu(thp)_2(dien)]$	X-ray	89
			$[Cu(thp)_2(CH_3NH_2)_2]$	ir	92
			$[Cu(thp)_2(C_2H_5NH_2)_2]$	ir	76
			$[Cu(thp)_2(C_3H_7NH_2)_2]$	ir	76
			$[Cu(thp)_2(C_4H_9NH_2)_2]$	ir	76
			$[Cu(thp)_2(C_6H_5NH_2)_2]$	ir	76
		Fe(II)	$[Fe(thp)_2(ClO_4)(H_2O)_2]ClO_4$	ir	105
				ir, cond, mag, spec	208
		Fe(III)	$[Fe(thp)_2(ClO_4)_2]ClO_4$	ir, cond, mag, spec	208
		Hg(I)	$[Hg(thp)Cl]$	ir, nmr, therm	70
				ir, therm	63
			$[Hg(thp)Cl_2]$	ir, nmr, therm	70
			$[Hg_2(thp)NO_3]$	ir	63
		Hg(II)	$[Hg(thp)Cl_2]$	ir	63
			$[(MeHg)(thp)]$	X-ray	64
		Mn(II)	$[Mn(thp)_2(ClO_4)_2]$	ir	105
				ir, cond, mag, spec	208
		Ni(II)	$[Ni(thp)_2(ClO_4)_2(H_2O)_2]$	ir	105
				cond, mag, ir, spec	208
		Pd(II)	$[Pd(thp)_2Br_2]$	ir, therm	209
			$[Pd(thp)_2Br_4]$	ir, therm	209
			$[Pd(thp)_2Cl_2]$	ir, nmr, therm	81
				ir, nmr	88
			$[Pd(thp)_2Cl_4]$	ir, nmr, therm	70
				ir, nmr	88
			$[Pd(thp)_2Br_4]$	ir, therm	209
		Pt(II)	$[Pt(thp)Cl_3]^-$	X-ray	74
			$[Pt(thp)_2Cl_2]$	uv, XPS	126
		Rh(II)	$[Rh_2(thp)_2(ac)_4]$	X-ray	66
		Ru(III)	$[Ru(NH_3)_5(thp)]^{2+}$	HPLC	127
			$[Ru(NH_3)_5(thp)]Cl_3$	uv, ech, pk	166
			$[Ru(NH_3)_4(thp)Cl]Cl_2$	uv, ech, pk	166
			$[Ru(NH_3)_4(thp)_2]Cl_3$	uv, ech, pk	166
		Ti(III)	$[(C_5H_5)_2Ti(thp)]$	X-ray, esr, ms	62

TABLE 1 (CONTINUED)

Base no.	Base	Metal	Stoichiometry	Method	Ref.
		Zn(II)	$[Zn(thp)_2(NH_3)_2]$	ir, spec, ks	59
				ir, therm	83
				ir	92
				ir	111
			$[Zn(thp)_2(ClO_4)_2]$	ir	105
				cond, ir, spec, mag	208
			$[Zn(thp)_2(en)]$	ir	111
			8-Ethyltheophylline		
X11	8etthp	Ag(I)	[Ag(8etthp)]	therm, pot, pk	69
		Pd(II)	$[Pd(8etthp)Br_3]$	ir, therm	209
			$[Pd(8etthp)_2Br_2]$	ir, therm	209
			$[Pd(8etthp)_2Cl_2]$	ir, therm	109
			7-Methyltheophylline		
X12	7methp	Pt(II)	$[Pt(7methp)_2Cl_2]$	XPS	126
			8-Pentyltheophylline		
X13	8pethp	Pd(II)	$[Pd(8pethp)Br_3]$	ir, therm	209
			$[Pd(8pethp)_2Br_2]$	ir, therm	209
			$[Pd(8pethp)_2Cl_2]$	ir, therm	109
			8-Phenyltheophylline		
X14	8phthp	Ag(I)	[Ag(8phthp)]	therm, pot, pk	69
			8-Propyltheophylline		
X15	8prthp	Ag(I)	$[Ag(8prthp)_2]$	therm, pot, pk	69
		Pd(II)	$[Pd(8prthp)Br_3]$	ir, therm	209
			$[Pd(8prthp)_2Br_2]$	ir, therm	209
			$[Pd(8prthp)_2Cl_2]$	ir, therm	109
			8-Isopropyltheophylline		
X16	8iprthp	Ag(I)	$[Ag(8iprthp)_2]$	therm, pot, pk	69
		Pd(II)	$[Pd(8iprthp)Br_3]$	ir, therm	209
			$[Pd(8iprthp)_2Br_2]$	ir, therm	209
			$[Pd(8iprthp)_2Cl_2]$	ir, therm	109

TABLE 1 (CONTINUED)

Base no.	Base	Metal	Stoichiometry	Method	Ref.
			Xanthine		
X1	xan	Ag(I)	$[Ag(xan)_2]NO_3$	nmr, therm	57
		Cd(II)	$[Cd(xan)_2]$	ir, therm, nmr	84
			$[Cd(xan)_2Cl_4]$	ir, therm, nmr	80
		Cr(III)	$[Cr(xan)_4]ClO_4$	ir, spec, mag	54
				ir, spec, mag	101
		Co(II)	$[Co(xan)_2]$	ir, therm, nmr	84
			$[Co(xan)_3]ClO_4$	ir, spec, mag	54
			$[Co(xan)_3]ClO_4.2EtOH$	ir, spec, mag	101
		Co(III)	$[Co(Bu_3P)(xan)(dmg)_2]$	nmr, X-ray	96
		Cu(II)	Cu…nta…xan	ks	42
			$[Cu(xan)_2]$	ir, therm, nmr	84
			$[Cu(xan)_2.2EtOH$	ir, spec, mag	86
		Fe(II)	$[Fe(xan)_3]ClO_4$	ir, spec, mag	54
		Fe(III)	$[Fe(xan)_4]ClO_4$	ir, spec, mag	54
				ir, spec, mag	101
		Hg(II)	$[Hg(xan)_2Cl_2]Cl_2$	ir, nmr, therm	80
			$[(MeHg)_2(xan)]$	nmr	116
			$[(MeHg)_3(xan)]$	nmr	116
		Mn(II)	$[Mn(xan)_3]ClO_4$	ir, spec, mag	54
		Ni(II)	$[Ni(xan)_3]ClO_4$	ir, spec, mag	54
			$[Ni(xan)_3]ClO_4.2EtOH$	ir, spec, mag	101
		Pd(II)	$[Pd(xan)_2Cl]$	ir, spec	177
			$[Pd(xan)_2Cl_2]$	ir, nmr, therm	81
				ir, nmr	88
			$[Pd(xan)_2Cl_2]Cl_2$	ir, nmr	88
		Pt(IV)	$[Pt(xan)_2Cl_3]$	ir, spec	177
		Ti(III)	$[((C_5H_5)_2Ti)_3Cl(xan)]$	X-ray	52
		Zn(II)	$[Zn(xan)_2Cl_4]$	ir, nmr, therm	80
			$[Zn(xan)_3]ClO_4$	ir, spec	54
			$[Zn(xan)_3]ClO_4.2EtOH$	ir	101
			8-Ethylxanthine		
X2	8etxan	Ag(I)	[Ag(8etxan)]	ir, nmr, spec	91
				ir, therm, X-ray	110
		Cu(II)	$[Cu(8etxan)_2(NH_3)_2]$	ir, nmr, spec	91
				ir, therm X-ray	110
		Pd(II)	$[Pd(8etxan)_2Br_2]$	therm, ir	209
			$[Pd(8etxan)_2Cl_2]$	ir, nmr, spec	91
				ir, therm, X-ray	110

TABLE 1 (CONTINUED)

Base no.	Base	Metal	Stoichiometry	Method	Ref.
			8-Ethyl-1-methylxanthine		
X3	8etlme-xan	Pd(II)	[Pd(8etlmexan)$_2$Br$_2$]	ir, therm	209
			8-Ethyl-3-methylxanthine		
X4	8et3-mexan	Ag(I)	[Ag(8et3mexan)]	ir, nmr, spec	91
				ir, therm, X-ray	110
		Au(III)	[Au(8et3mexan)Cl$_4$]	ir, nmr, spec	91
				ir, therm, X-ray	110
		Cu(II)	[Cu(8et3mexan)$_2$(NH$_3$)$_2$]	ir, nmr, spec	91
				ir, therm, X-ray	110
		Pd(II)	[Pd(8et3mexan)$_2$Cl$_2$]	ir, nmr, spec	91
				ir, therm, X-ray	110
			1-Methylxanthine		
X5	1mexan	Pd(II)	[Pd(NH$_3$)$_2$(1mexan)$_2$]	ir, XPS	103
		Rh(III)	[Rh(1mexan)$_3$Cl$_3$]	ir, XPS	103
			3-Methylxanthine		
X6	3mexan	Pd(II)	[Pd(3mexan)$_2$Cl$_2$]	ir	103
		Pt(II)	[Pt(3mexan)$_2$(NH$_3$)$_2$]	ir	103
				therm	97
			7-Methylxanthine		
X7	7me-xan	Pd(II)	[Pd(7mexan)$_2$]Cl$_2$	ir, XPS	103
			[Pd(NH$_3$)$_2$(7mexan)$_2$]	ir, XPS	103
			[Pd(NH$_3$)$_2$(7mexan)]$_2$	ir, XPS	103
			8-Methylxanthine		
X8	8me-xan	Pd(II)	[Pd(NH$_3$)$_2$(8mexan)$_2$]	ir, XPS	103
			[Pd(NH$_3$)$_2$(8mexan)]$_2$	ir, XPS	103
			9-Methylxanthine		
X9	9me-xan	Pd(II)	[Pd(NH$_3$)$_2$(9mexan)$_2$]	therm	97
				therm, ir, XPS	103
			[Pd(NH$_3$)$_2$(9mexan)]$_2$	ir	103

TABLE 1 (CONTINUED)

Base no.	Base	Metal	Stoichiometry	Method	Ref.
		Pt(II)	$[Pt(NH_3)_2(9mexan)]Cl$	ir, therm	103
				ir	148
			$[Pt(NH_3)_2(9mexan)_2]$	ir	103
			1,3-Dimethylxanthine		
X10	thp	see Theophylline			
			1,9-Dimethylxanthine		
X19	19dme-xan	Ru(III)	$[Ru(NH_3)_5(19dmexan)]Cl_3$	pk, uv, ech	166
			3,7-Dimethylxanthine		
X20	thb	see Theobromine			
			3,8-Dimethylxanthine		
X25	38dme-xan	Ag(I)	$[Ag(38dmexan)]$	ir, uv, nmr pk	60
				ir, nmr, therm	68
		Au(III)	$[Au(38dmexan)Cl_4]$	ir, nmr, mag, therm	82
		Cu(II)	$[Cu(38dmexan)_2Cl_4]$	therm, mag	58
		Hg(II)	$[Hg_2(38dmexan)NO_3]$	ir, nmr, therm	68
		Pd(II)	$[Pd(38dmexan)_2Br_2]$	ir, therm	209
			$[Pd(38dmexan)_2Cl_2]$	ir, nmr, therm	81
				ir, nmr	88
			3,9-Dimethylxanthine		
X26	39dme-xan	Ru(III)	$[Ru(39dmexan)(NH_3)_5Cl_3]$	pk, uv, ech	166
			1,3,7-Trimethylxanthine		
X30	caf	see Caffeine			
			1,3,8-Trimethylxanthine		
X35	8tme-xan	Ag(I)	$[Ag(8tmexan)]$	ir, uv, nmr, pk	60
				ir, nmr, therm	68
		Au(III)	$[Au(8tmexan)Cl_4]$	ir, nmr, therm, mag	82
		Cd(II)	$[Cd(8tmexan)_2(NH_3)_2]$	ir, nmr, therm	87

TABLE 1 (CONTINUED)

Base no.	Base	Metal	Stoichiometry	Method	Ref.
		Cu(II)	[Cu(8tmexan)$NH_3)_2$]	ir, nmr, mag, therm	87
		Hg(I)	[Hg_2(8tmexan)NO_3]	ir, nmr, therm	81
		Hg(II)	[Hg(8tmexan)Cl_2]	ir, nmr, therm	87
		Pd(II)	[Pd(8tmexan)Br_3]	ir, therm	209
			[Pd$(8tmexan)_2Br_2$]	ir, therm	209
			[Pd$(8tmexan)_2Cl_2$]	ir, nmr, therm	81
				ir, nmr	88
		Zn(II)	[Zn(8tmexan)$NH_3)_2$]	ir, nmr, therm	87
			1,3,9-Trimethylxanthine		
X36	9tme-xan	Pt(II)	[Pt(en)(9tmexan)]$NO_3)_2$	X-ray, nmr	113
			[Pt(en)(9tmexan)]$PF_6)_2$	X-ray, nmr	113
		Ru(III)	[Ru$(NH_3)_5$(9tmexan)]Cl_3	pk, ech, uv	166
			2-Thioxanthine		
X37	2Sxan	Cd(II)	Cd…2Sxan	pot, pk	108
		Zn(II)	Zn…2Sxan	pot, pk	108
			6-Thioxanthine		
X38	6Sxan	Pt(II)	[Pt(6Sxan)$(NH_3)_2$]	atta, uv, ir, OR	65

REFERENCES

1. **R. Beyerle-Pfnur, B. Brown, R. Faggiani, B. Lippert, and J. L. Lock,** *Inorg. Chem.*, 1985, *24*, 4001.
2. **E. Sletten, T. Marthinsen, and J. Sletten,** *Inorg. Chim. Acta,* 1985, *106*, 1.
3. **E. Sletten, T. Marthinsen, and J. Sletten,** *Inorg. Chim. Acta,* 1984, *93*, 37.
4. **R. Beyerle and B. Lippert,** *Inorg. Chim. Acta,* 1982, *66*, 141.
5. **A. N. Speca, C. M. Mikulski, F. J. Iaconianni, L. L. Pytlewski, and N. M. Karayannis,** *Inorg. Chim. Acta,* 1979, *37*, L551.
6. **W. S. Sheldrick and P. Bell,** *Inorg. Chim. Acta,* 1986, *123*, 181.
7. **T. Sorrell, L. A. Epps, T. J. Kistenmacher, and L. G. Marzilli,** *J. Am. Chem. Soc.*, 1978, *100*, 5756.
8. **A. A. Zaki, C. A. McAuliffe, M. E. Friedman, W. E. Hill, and H. H. Kohl,** *Inorg. Chim. Acta,* 1983, *69*, 93.
9. **T. R. Harkins and H. Freiser,** *J. Am. Chem. Soc.*, 1958, *80*, 1132.
10. **M. M. T. Khan, S. Satyanarayana, M. S. Jyoti, and C.A. Lincoln,** *Indian J. Chem. Sect. A,* 1983, *22A*, 357.
11. **J. -P. Charland, J. F. Britten, and A. L. Beauchamp,** *Inorg. Chim. Acta,* 1986, *124*, 161.
12. **J. D. Orbell, C. Solorzano, L. G. Marzilli, and T. J. Kistenmacher,** *Inorg. Chem.*, 1982, *21*, 2630.
13. **C. T. Mortimer, B. Miller, and M. P. Wilkinson,** *Inorg. Chim. Acta,* 1980, *46*, 285.
14. **M. E. Kastner, K. F. Coffey, M. J. Clarke, S. E. Edmonds, and K. Eriks,** *J. Am. Chem. Soc.,* 1981, *103*, 5747.
15. **I. I. Volchenskova, N. N. Maidanevich, and L. I. Budarin,** *Inorg. Chim. Acta,* 1983, *79*, 246.
16. **B. T. Khan, S. V. Kumari, and G. N. Goud,** *J. Coord. Chem.* 1982, *12*, 19.
17. **B. T. Khan, S. V. Kumari, and G. N. Goud,** *Indian J. Chem. Sect. A,* 1982, *21A*, 264.
18. **B. T. Khan and A. Mehmood,** *J. Inorg. Nucl. Chem.*, 1978, *40*, 1938.
19. **M. Poojary and H. Manohar,** *Inorg. Chim. Acta,* 1984, *93*, 153.
20. **N. Farrell and N. G. de Oliveira,** *Inorg. Chim. Acta,* 1982, *66*, L61.
21. **C. M. Mikulski, D. Delacato, B. Braccia, and N. M. Karayannis,** *Inorg. Chim. Acta,* 1984, *93*, L19.
22. **C. M. Mikulski, S. Cocco, N. de Franco, and N. M. Karayannis,** *Inorg. Chim. Acta,* 1982, *67*, 61.
23. **R. G. Bhattacharyya and I. Bhaduri,** *J. Indian Chem. Soc.,* 1982, *59*, 919.
24. **C. M. Mikulski, S. Cocco, N. de Franco, and N. M. Karayannis,** *Inorg. Chim. Acta,* 1983, *78*, L25.
25. **C. M. Mikulski, S. Cocco, N. de Franco, and N. M. Karayannis,** *Inorg. Chim. Acta,* 1983, *80*, L23.
26. **C. M. Mikulski, S. Cocco, N. de Franco, and N. M. Karayannis,** *Inorg. Chim. Acta,* 1983, *80*, L71.
27. **C. M. Mikulski, S. Cocco, N. de Franco, and N. M. Karayannis,** *Inorg. Chim. Acta,* 1983, *80*, L61.
28. **A. Terzis,** *Inorg. Chem.,* 1976, *15*, 793.
29. **C. M. Mikulski, R. de Prince, T. B. Tran, F. J. Iaconianni, L. L. Pytlewski, A. N. Speca, and N. M. Karayannis,** *Inorg. Chim. Acta,* 1981, *56*, 163.
30. **T. Theophanides, M. Berjot, and L. Bernard,** *J. Raman Spectrosc.,* 1977, *6*, 109.
31. **M. A. Guichelaar and J. Reedijk,** *Recl. Trav. Chim. Pays-Bas,* 1978, *97*, 295.
32. **L. Prizant, M. J. Olivier, R. Rivest, and A. L. Beauchamp,** *Can. J. Chem.,* 1981, *59*, 1311.
33. **A. Terzis, N. Hadjiliadis, R. Rivest, and T. Theophanides,** *Inorg. Chim. Acta,* 1975, *12*, L5.

34. **C. M. Mikulski, D. Braccia, D. Delacato, J. Fleming, D. Fleming, and N. M. Karayannis,** *Inorg. Chim. Acta,* 1985, *106*, L13.
35. **R. Beyerle-Pfnur, S. Jaworski, B. Lippert, H. Schollhorn, and U. Thewalt,** *Inorg. Chim. Acta,* 1985, *107*, 217.
36. **R. Sridharan and C. R. Krishnamoorthy,** *J. Coord. Chem.,* 1983, *12*, 231.
37. **J. -P. Charland, M. Simard, and A. L. Beauchamp,** *Inorg. Chim. Acta,* 1983, *80*, L57.
38. **M. J. Olivier and A. L. Beauchamp,** *Inorg. Chem.,* 1980, *19*, 1064.
39. **A. N. Speca, L. L. Pytlewski, C. M. Mikulski, and N. M. Karayannis,** *Inorg. Chim. Acta,* 1982, *66*, L53.
40. **A. B. Robins,** *Chem. Biol. Interact.,* 1973, *6*, 35.
41. **T. Beringhelli, M. Freni, F. Morazzoni, P. Romiti, and R. Servida,** *Spectrochim. Acta Part A,* 1981, *37A*, 763.
42. **R. Ghose and K. Dey,** *Acta Chim. Acad. Sci. Hung.,* 1981, *108*, 9.
43. **D. Camboli, J. Besancon, J. Tirouflet, B. Gautheron, and P. Meunier,** *Inorg. Chim. Acta,* 1983, *78*, L51.
44. **D. W. Abbott and C. Woods,** *Inorg. Chem.,* 1983, *22*, 597.
45. **C. H. Wei and K. B. Jacobson,** *Inorg. Chem.,* 1981, *20*, 356.
46. **J. -P. Macquet and T. Theophanides,** *Inorg. Chim. Acta,* 1976, *18*, 189.
47. **J. -P. Macquet and T. Theophanides,** *Biopolymers,* 1975, *14*, 281.
48. **A. I. Stetsenko and E. S. Dmitriyeva,** *Koord. Khim.,* 1977, *3*, 1240.
49. **A. I. Stetsenko, E. S. Dmitriyeva, and K. I. Yakovlev,** *J. Clin. Hematol. Oncol.,* 1977, *7*, 522.
50. **C. M. Mikulski, L. Mattucci, L. Weiss, and N. M. Karayannis,** *Inorg. Chim. Acta,* 1985, *107*, 147.
51. **L. G. Marzilli, K. Wilkowski, C. C. Chiang, and T. J. Kistenmacher,** *J. Am. Chem. Soc.,* 1979, *101*, 7504.
52. **A. L. Beauchamp, F. Belanger-Gariepy, A. Mardhy, and D. Cozak,** *Inorg. Chim. Acta,* 1986, *124*, L23.
53. **M. A. Romero-Molina, E. Colacio-Rodriguez, J. Ruis-Sanchez, J. M. Salas-Peregrin, and F. Nieto,** *Inorg. Chim. Acta,* 1986, *123*, 133.
54. **C. M. Mikulski, M. K. Kurlan, M. Bayne, M. Gaul, and N. M. Karayannis,** *Inorg. Chim. Acta,* 1986, *123*, 27.
55. **L. G. Marzilli, T. J. Kistenmacher, and C. -H. Chang,** *J. Am. Chem. Soc.,* 1971, *93*, 2736.
56. **G. Y. H. Chu, S. Mansy, R. E. Duncan, and R. S. Tobias,** *J. Am. Chem. Soc.* 1978, *100*, 593.
57. **E. Colacio-Rodriguez, J. M. Salas-Peregrin, J. D. Lopez-Gonzalez, and C. V. Calahorro,** *An. Quim.,* 1984, *80B*, 49.
58. **E. Colacio-Rodriguez, J. M. Salas-Peregrin, and J. D. Lopez Gonzalez,** *An. Quim.,* 1984, *80B*, 223.
59. **J. M. Salas-Peregrin, E. Colacio-Rodriguez, M. M. Carretero, and J. D. Lopez-Gonzalez,** *An. Quim.,* 1984, *80B*, 167.
60. **E. Colacio-Rodriguez, J. M. Salas-Peregrin, M. N. Montiel, and A. Sanchez Rodrigo,** *An. Quim.,* 1984, *80B*, 441.
61. **B. T. Khan, S. V. Kumari, K. M. Mohan, and G. Narsa Goud,** *Polyhedron,* 1985, *4*, 1617.
62. **D. Cozak, A. Mardhy, M. J. Olivier, and A. L. Beauchamp,** *Inorg. Chem.,* 1986, *25*, 2600.
63. **E. Colacio-Rodriguez, J. M. Salas-Peregrin, and M. A. Romero-Molina,** *Rev. Chim. Miner.,* 1984, *21*, 123.
64. **A. R. Norris, S. E. Taylor, E. Buncel, F. Belanger-Gariepy, and A. L. Beauchamp,** *Inorg. Chim. Acta,* 1984, *92*, 271.

65. **A. Maeda, N. Abiko, and T. Sasaki,** *J. Pharm. Dyn.,* 1982, *5*, 81.
66. **K. Aoki and H. Yamazaki,** *J. Chem. Soc. Chem. Commun.,* 1980, 186.
67. **D. M. L. Goodgame, P. B. Hayman, R. T. Riley, and D. J. Williams,** *Inorg. Chim. Acta,* 1984, *91*, 89.
68. **E. Colacio-Rodriguez, J. M. Salas-Peregrin, M. A. Romero-Molina, and R. Lopez-Garzon,** *Thermochim. Acta,* 1984, *76*, 373.
69. **J. M. Salas-Peregrin, E. Colacio-Rodriguez, F. Girela-Vichez, and M. Roldan-Medina,** *Thermochim. Acta,* 1984, *80*, 323.
70. **E. Colacio-Rodriguez and J. M. Salas-Peregrin,** *Thermochim. Acta,* 1984, *74*, 45.
71. **T. J. Kistenmacher, K. Wilkowski, B. de Castro, C. C. Chiang, and L. G. Marzilli,** *Biochem. Biophys. Res. Commun.,* 1979, *91*, 1521.
72. **L. Y. Kuo, M. G. Kanatzidis, and T. J. Marks,** *J. Am. Chem. Soc.,* 1987, *109*, 7207.
73. **N. P. Johnson, A. M. Mazard, J. Escalier, and J. P. Marquet,** *J. Am. Chem. Soc.,* 1985, *107*, 6376.
74. **E. H. Griffith and E. L. Amma,** *J. Chem. Soc. Chem. Commun.,* 1979, 322.
75. **M. B. Cingi, A. M. M. Lanfredi, A. Tiripiochio, and M. T. Camellini,** *Transition Met. Chem.,* 1979, *4*, 221.
76. **W. J. Birdsall and M. S. Zitzman,** *J. Inorg. Nucl. Chem.,* 1979, *41*, 116.
77. **C. M. Mikulski, T. B. Tran, L. Mattucci, and N. M. Karayannis,** *Inorg. Chim. Acta,* 1983, *78*, 269.
78. **J. Arpalahti and H. Lonnberg,** *Inorg. Chim. Acta,* 1983, *80*, 25.
79. **J. Arpalahti and H. Lonnberg,** *Inorg. Chim. Acta,* 1983, *78*, 63.
80. **E. Colacio-Rodriguez, J. D. Lopez-Gonzalez, and J. M. Salas-Peregrin,** *J. Therm. Anal.,* 1983, *28*, 3.
81. **J. M. Salas-Peregrin, E. Colacio-Rodriguez, M. A. Romero-Molina, and M. P. Sanchez-Sanchez,** *Thermochim. Acta,* 1983, *69*, 313.
82. **E. Colacio-Rodriguez, J. M. Salas-Peregrin, R. Lopez-Garzon, and J. D. Lopez-Gonzalez,** *Thermochim. Acta,* 1983, *71*, 139.
83. **E. Colacio-Rodriguez, J. M. Salas-Peregrin, M. P. Sanchez-Sanchez, and A. Mata-Arjona,** *Thermochim. Acta,* 1983, *66*, 245.
84. **J. M. Salas-Peregrin, E. Colacio-Rodriguez, J. D. Lopez-Gonzalez, and C. Valenzuela-Calahorro,** *Thermochim. Acta,* 1983, *63*, 145.
85. **A. T. M. Marcelis, J. L. Van der Veer, J. C. M. Zwetsloot, and J. Reedijk,** *Inorg. Chim. Acta,* 1983, *78*, 195.
86. **C. M. Mikulski, T. B. Tran, L. Mattucci, and N. M. Karayannis,** *Inorg. Chim. Acta,* 1983, *78*, 211.
87. **E. Colacio-Rodriguez, J. D. Lopez-Gonzalez, and J. M. Salas-Peregrin,** *Can. J. Chem.,* 1983, *61*, 2506.
88. **E. Colacio, J. M. Salas, M. A. Romero, A. Sanchez, and M. Nogueras,** *Inorg. Chim. Acta,* 1983, *79*, 250.
89. **T. Sorrell, L. G. Marzilli, and T. J. Kistenmacher,** *J. Am. Chem. Soc.,* 1976, *98*, 2181.
90. **A. N. Speca, C. M. Mikulski, F. J. Iaconianni, L. L. Pytlewski, and N. M. Karayannis,** *Inorg. Chim. Acta,* 1980, *46*, 235.
91. **J. M. Salas-Peregrin, E. Sanchez-Martinez, and E. Colacio-Rodriguez,** *Inorg. Chim. Acta,* 1985, *107*, 23.
92. **M. S. Zitzman, R. R. Krebs, and W. J. Birdsall,** *J. Inorg. Nucl. Chem.,* 1978, *40*, 571.
93. **N. B. Behrens, B. A. Cartwright, D. M. L. Goodgame, and A. C. Skapski,** *Inorg. Chim. Acta,* 1978, *31*, L471.
94. **C. M. Mikulski, R. de Prince, T. B. Tran, and N. M. Karayannis,** *Inorg. Chim. Acta,* 1981, *56*, 27.
95. **H. J. Krentzien, M. J. Clarke, and H. Taube,** *Bioinorg. Chem.,* 1975, *4*, 143.
96. **L. G. Marzilli, L. A. Epps, T. Sorrell, and T. J. Kistenmacher,** *J. Am. Chem. Soc.,* 1975, *97*, 3351.

97. **H. S. O. Chan and J. R. Lusty,** *J. Therm. Anal.,* 1985, *30*, 25.
98. **R. Cini, A. Cinquantini, M. Sabat, and L. G. Marzilli,** *Inorg. Chem.,* 1985, *24*, 3903.
99. **J. H. J. den Hartog, M. L. Salm, and J. Reedijk,** *Inorg. Chem.,* 1984, *23*, 2001.
100. **B. L. Heyl, K. Shinozuka, S. K. Miller, and D. G. van der Veer,** *Inorg. Chem.,* 1985, *24*, 661.
101. **C. M. Mikulski, M. Kurlan, and N. M. Karayannis,** *Inorg. Chim. Acta,* 1985, *106*, L25.
102. **Y. Wei-Da, L. Mei-Qing, and P. Shi-Qi,** *Inorg. Chim. Acta,* 1985, *106*, 65.
103. **J. R. Lusty, H. S. O. Chan, E. Khor, and J. Peeling,** *Inorg. Chim. Acta,* 1985, *106*, 209.
104. **A. Grigoratos and N. Katsaros,** *Inorg. Chim. Acta,* 1985, *108*, 41.
105. **C. M. Mikulski, M. K. Kurlan, S. Grossman, M. Bayne, and N. M. Karayannis,** *Inorg. Chim. Acta,* 1985, *108*, L7
106. **J. Baranowska-Kortylewicz, E. J. Pavlik, W. T. Smith, R. C. Flanigan, J. R. van Nagell, D. Ross, and D. E. Kenady,** *Inorg. Chim. Acta,* 1985, *108*, 91.
107. **L. G. Marzilli, T. J. Kistenmacher, and C. -H. Chiang,** *J. Am. Chem. Soc.,* 1973, *95*, 7507.
108. **M. P. Sanchez-Sanchez, J. M. Salas-Peregrin, M. A. Romero-Molina, and A. Mata-Arjona,** *Thermochim. Acta,* 1985, *88*, 355.
109. **E. Colacio-Rodriguez, J. M. Salas-Peregrin, J. Ruiz-Sanchez, and E. Garcia-Mejias,** *Thermochim. Acta,* 1985, *89*, 159.
110. **J. M. Salas-Peregrin, E. Colacio-Rodriguez, and E. Sanchez-Martinez,** *Thermochim. Acta,* 1985, *86*, 189.
111. **W. J. Birdsall,** *Inorg. Chim. Acta,* 1985, *99*, 59.
112. **A. T. M. Marcelis, H. -J. Korte, B. Krebs, and J. Reedijk,** *Inorg. Chem.,* 1982, *21*, 4059.
113. **J. D. Orbell, K. Wilkowski, B. de Castro, L. G. Marzilli, and T. J. Kistenmacher,** *Inorg. Chem.,* 1982, *21*, 813.
114. **T. J. Kistenmacher, B. de Castro, K. Wilkowski, and L. G. Marzilli,** *J. Inorg. Biochem.,* 1982, *16*, 33.
115. **C. R. Krishnamoorthy and M. M. T. Khan,** *J. Coord. Chem.,* 1983, *12*, 313.
116. **A. R. Norris and R. Kumar,** *Inorg. Chim. Acta,* 1984, *93*, L63.
117. **P. J. Toscano, C. C. Chiang, T. J. Kistenmacher, and L. G. Marzilli,** *Inorg. Chem.,* 1981, *20*, 1513
118. **B. de Castro, C. C. Chiang, K. Wilkowski, L. G. Marzilli, and T. J. Kistenmacher,** *Inorg. Chem.,* 1981, *20*, 1835.
119. **D. Sonnenfroh and R. W. Kreilick,** *Inorg. Chem.,* 1980, *19*, 1259.
120. **A. N. Speca, C. M. Mikulski, F. J. Iaconianni, L. L. Pytlewski, and N. M. Karayannis,** *Inorg. Chem.,* 1980, *19*, 3491.
121. **W. S. Sheldrick,** *Z. Naturforsch. Teil B.,* 1982, *37B*, 653.
122. **G. Pneumatikakis,** *Inorg. Chim. Acta,* 1984, *93*, 5.
123. **J. Arpalahti and E. Ottoila,** *Inorg. Chim. Acta,* 1985, *107*, 105.
124. **J. Arpalahti and H. Lonnberg,** *Inorg, Chim. Acta,* 1985, *107*, 197.
125. **A. L. Beauchamp, D. Cozak, and A. Mardhy,** *Inorg. Chim. Acta,* 1984, *92*, 191.
126. **P. Umapathy and R. A. Harnesswala,** *Polyhedron,* 1983, *2*, 129.
127. **M. J. Clarke, D. F. Coffey, H. J. Perpall, and J. Lyon,** *Anal. Biochem.,* 1982, *122*, 404.
128. **W. S. Sheldrick,** *Acta Crystallogr. Sect. B,* 1981, *B37*, 945.
129. **P. I. Vestues and E. Sletten,** *Inorg. Chim. Acta,* 1981, *52*, 269.
130. **T. A. Connors and J. J. Roberts, Eds.,** *2nd Int. Symp. on Platinum Coordination Complexes in Cancer Chemotherapy,* Springer-Verlag, Berlin, 1974.
131. 3rd International Symp. on Platinum Coordination Complexes in Cancer Chemotherapy, *J. Clin. Hematol. Oncol.,* 1977, 7.
132. **A. P. Hitchcock, C. J. L. Lock, and W. M. C. Pratt,** in *Platinum, Gold, and Other Metal Chemotherapeutic Agents* (ACS Symp. 209), S. J. Lippard, Ed., American Chemical Society, Washington, D.C., 1983.

133. **M. P. Hacker, E. B. Douple, and I. H. Krakoff, Eds.,** *4th Int. Symp. on Platinum Coordination Complexes in Cancer Chemotherapy,* Martinus Nijhoff, Boston, 1984.
134. **B. Beyerle-Pfnur and B. Lippert,** in *4th Int. Symp. on Platinum Coordination Complexes in Cancer Chemotherapy,* M. P. Hacker, E. B. Douple, and I. H. Krakoff, Eds., Martinus Nijhoff, Boston, 1984, 53.
135. **G. Raudaschl and B. Lippert,** in *4th Int. Symp. on Platinum Coordination Complexes in Cancer Chemotherapy,* M. P. Hacker, E. B. Douple, and I. H. Krakoff, Eds., Martinus Nijhoff, Boston, 1984, 54.
136. **M. Nicolini, Ed.,** *5th Int. Symp. on Platinum and Other Metal Coordination Compounds in Cancer Chemotherapy,* Martinus Nijhoff, Boston, 1988.
137. **B. Lippert, J. Arpalahti, O. Krizanovic, W. Micklitz, F. Schwartz, and G. Trotscher,** in *5th Int. Symp. on Platinum and Other Metal Coordination Compounds in Cancer Chemotherapy,* M. Nicolini, Ed., Martinus Nijhoff, Boston, 1988, 563.
138. **L. G. Marzilli,** in *Advances in Inorganic Biochemistry, Vol. 3,* Metal Ions in Genetic Information Transfer, G. L. Eichhorn and L. G. Marzilli, Eds., Elsevier, New York, 1981, 47.
139. **H. Sigel, Ed.,** *Metal Ions in Biological Systems,* Marcel Dekker, New York.
140. **M. J. Cleare,** *Coord. Chem. Rev.,* 1974, *12*, 349.
141. **J. Dehand and J. Jordanov,** *J. Chem. Soc. Dalton Trans.,* 1977, 2588.
142. **J. J. Roberts,** in *Advances in Inorganic Biochemistry, Vol. 3,* Metal Ions in Genetic Information Transfer, G. L. Eichhorn and L. G. Marzilli, Eds., Elsevier, New York, 1981, 274.
143. **F. R. Hartley,** *Coord. Chem. Rev.,* 1985, *67*, 1.
144. **D. R. Williams,** *Chem. Rev.,* 1972, *72*, 203.
145. **C. M. Mikulski, R. de Prince, G. W. Madison, M. Gaul, and N. M. Karayannis,** *Inorg. Chim. Acta,* 1987, *138*, 55.
146. **E. Dubler and E. Gyr,** *Inorg. Chem.,* 1988, *27*, 1466.
147. **T. Theophanides,** *Can. J. Spectrosc.,* 1981, *26*, 165.
148. **J. R. Lusty and P. F. Lee,** *Inorg. Chim. Acta,* 1984, *91*, L47.
149. **L. G. Marzilli, T. J. Kistenmacher, P. E. Darcy, D. J. Szalda, and N. Beer,** *J. Am. Chem. Soc.,* 1974, *96*, 4688.
150. **G. Raudaschl-Sieber, H. Schollhorn, U. Thewalt, and B. Lippert,** *J. Am. Chem. Soc.,* 1985, *107*, 3591.
151. **G. Raudaschl-Sieber, L. G. Marzilli, B. Lippert, and K. Shinozuka,** *Inorg. Chem.,* 1985, *24*, 989.
152. **R. Faggiani, B. Lippert, C. J. L. Lock, and R. A. Speranzini,** *Inorg. Chem.,* 1982, *21*, 3216.
153. **S. Mansy, B. Rosenberg, and A. J. Thomson,** *J. Am. Chem. Soc.,* 1973, *95*, 1633.
154. **B. Lippert,** *J. Am. Chem. Soc.,* 1981, *103*, 5691.
155. **C. M. Mikulski, L. Mattucci, Y. Smith, T. B. Tran, and N. M. Karayannis,** *Inorg. Chim. Acta,* 1983, *80*, 127.
156. **C. M. Mikulski, L. Mattucci, L. Weiss, and N. M. Karayannis,** *Inorg. Chim. Acta,* 1984, *92*, 181.
157. **C. M. Mikulski, L. Mattucci, L. Weiss, and N. M. Karayannis,** *Inorg. Chim. Acta,* 1984, *92*, L29.
158. **J. D. Orbell, K. Wilkowski, L. G. Marzilli, and T. J. Kistenmacher,** *Inorg. Chem.,* 1982, *21*, 3478.
159. **C. M. Mikulski, L. Mattucci, L. Weiss, and N. M. Karayannis,** *Inorg. Chim. Acta,* 1985, *108*, L35.
160. **C. M. Mikulski, L. Mattucci, L. Weiss, and N. M. Karayannis,** *Inorg. Chim. Acta,* 1984, *92*, 275.
161. **H. Schollhorn, G. Raudaschl-Sieber, G. Muller, U. Thewalt, and B. Lippert,** *J. Am. Chem. Soc.,* 1985, *107*, 5932.

162. **C. M. Mikulski, L. Mattucci, L. Weiss, and N. M. Karayannis,** *Inorg. Chim. Acta,* 1985, *107,* 81.
163. **J. D. Orbell, C. Solozano, L. G. Marzilli, and T. J. Kistenmacher,** *Inorg. Chem.,* 1982, *21,* 3806.
164. **M. J. Clarke and H. Taube,** *J. Am. Chem. Soc.,* 1974, *96,* 5413.
165. **J. Dehand and J. Jordanov,** *J. Chem. Soc. Chem. Commun.,* 1976, 594.
166. **M. J. Clarke and H. Taube,** *J. Am. Chem. Soc.,* 1975, *97,* 1397.
167. **J. L. van der Veer, H. van der Elst, and J. Reedijk,** *Inorg. Chem.,* 1987, *26,* 1536.
168. **B. Lippert, G. Raudaschl, C. J. L. Lock, and P. Pilon,** *Inorg. Chim. Acta,* 1984, *93,* 43.
169. **G. Raudaschal and B. Lippert,** *Inorg. Chim. Acta,* 1983, *80,* L49.
170. **D. J. Szalda, T. J. Kistenmacher, and L. G. Marzilli,** *J. Am. Chem. Soc.,* 1976, *98,* 8371.
171. **C. C. Chiang, L. A. Epps, L. G. Marzilli, and T. J. Kistenmacher,** *Inorg. Chem.,* 1979, *18,* 791.
172. **T. J. Kistenmacher, D. J. Szalda, C. C. Chiang, M. Rossi, and L. G. Marzilli,** *Inorg. Chem.,* 1978, *9,* 2582
173, **M. J. Clarke,** *Inorg. Chem.,* 1977, 16, 738.
174. **G. Raudaschl-Sieber and B. Lippert,** *Inorg. Chem,* 1985, *24,* 2426.
175. **T. Osa, H. Hino, S. Fujieda, T. Shiio, and T. Kono,** *Chem. Pharm. Bull.,* 1986, *34,* 3563.
176. **S. Murakami, K. Saito, A. Muro-Matsu, M. Moriyasu, A. Kato, and K. Hashimoto,** *Inorg. Chim. Acta,* 1988, *152,* 91.
177. **C. M. Mikulski, S. Grossman, C. J. Lee, and N. M. Karayannis,** *Transition Met. Chem.,* 1987, *12,* 21.
178. **M. A. Zoroddu, R. Dallocchio, and M. A. Cabras,** *Transition Met. Chem.,* 1987, *12,* 356.
179. **M. R. Caira and L. R. Nassimbeni,** *Acta Crystallogr. Sect. B,* 1975, *B31,* 1339.
180. **L. G. Purnell, E. D. Estes, and D. J. Hodgson,** *J. Am. Chem. Soc.,* 1976, *98,* 740.
181. **T. J. Kistenmacher, D. J. Szalda, and L. G. Merzilli,** *Inorg. Chem.,* 1975, *14,* 1686.
182. **D. J. Szalda, T. J. Kistenmacher, and L. G. Marzilli,** *Inorg. Chem.,* 1976, *15,* 2783.
183. **T. J. Kistenmacher,** *Acta Crystallogr. Sect. B,* 1975, *B31,* 85.
184. **T. J. Kistenmacher, D. J. Szalda, and L. G Marzilli,** *Acta Crystallogr. Sect. B,* 1975, *B31,* 2416.
185. **E. Sletten,** *J. Chem. Soc. Chem. Commun.,* 1971, 558.
186. **E. Sletten,** *Acta Crystallogr. Sect. B,* 1974, *B30,* 1961.
187. **E. Sletten,** *Acta Crystallogr. Sect. B,* 1970, *B26,* 1609.
188. **M. Sundaralingam and J. A. Carrabsine,** *J. Mol. Biol.,* 1971, *61,* 287.
189. **J. P. DeClerq, M. Debbaudt, and M. van Meersche,** *Bull. Soc. Chim. Belges,* 1971, *80,* 527.
190. **J. A. Carrabine and M. Sundaralingam,** *J. Am. Chem. Soc.,* 1970, *92,* 369.
191. **L. G. Marzilli,** *Prog. Inorg. Chem.,* 1977, *23,* 255.
192. **D. J. Hodgson,** *Prog. Inorg. Chem.,* 1977, *23,* 211.
193. **E. Sletten and N. Flogstad,** *Acta Crystallogr. Sect. B,* 1976, *B32,* 461.
194. **L. G. Purnell, J. C. Shepherd, and D. J. Hodgson,** *J. Am. Chem. Soc.,* 1975, *97,* 2376.
195. **M. Sundaralingam, C. D. Stout, and S. M. Hecht,** *J. Am. Chem. Soc., Chem. Commun.,* 1971, 240.
196. **C. D. Stout, M. Sundaralingam, and G. H. Y. Lin,** *Acta Crystallogr. Sect. B,* 1972, *B28,* 2136.
197. **L. Srinivasan and M. Taylor,** *J. Chem. Soc. Chem. Commun.,* 1970, 1668.
198. **M. R. Taylor,** *Acta Crystallogr. Sect. B,* 1973, *B29,* 884.
199. **P. de Meester and A. C. Shapski,** *J. Am. Chem. Soc. Dalton Trans.,* 1973, 424.
200. **M. Green,** *Transition Met. Chem.,* 1987, *12,* 186.
201. **P. Laverture, J. Hubert, and A. L. Beauchamps,** *Inorg. Chem.,* 1976, *15,* 322.
202. **M. A. Cabras and M. A. Zoroddu,** *Inorg. Chim. Acta,* 1987, *136,* 17.
203. **A. Fontana, F. Maggio, and T. Pizzino,** *J. Inorg. Biochem.,* 1987, *29,* 165.
204. **B. T. Khan, A. Gaffuri, N. P. Rao, and S. M. Zakeeruddin,** *Polyhedron,* 1987, *6,* 387.

205. **E. Dubler, G. Hauggi, and W. Bensch,** *J. Inorg. Biochem.,* 1987, *29*, 269.
206. **A. I. Stetsenko, O. M. Adamov, E. S. Dimitrieva, and M. V. Muratova,** *Koord. Khim.,* 1987, *13*, 377.
207. **J. L. van der Veer, J. G. Ligtvoet, and J. Reedijk,** *J. Inorg. Biochem.,* 1987, *29*, 217.
208. **C. M. Mikulski, M. K. Kurlan, S. Grossman, M. Bayne, M. Gaul, and N. M. Karayannis,** *J. Coord. Chem.,* 1987, *15*, 347.
209. **M. I. Moreno-Vida, E. Colacio-Rodriguez, M. N. Moreno-Carvetero, J. Ruiz-Sanchez, and J. M. Salas-Peregrin,** *Thermochim. Acta,* 1987, *115*, 45.
210. **E. L. J. Breet and R. van Eldik,** *Inorg. Chem.,* 1987, *26*, 2517.
211. **R. El-Mehdawi, S. A. Bryan, and D. M. Roundhill,** *J. Am. Chem. Soc.,* 1985, *107*, 6282.
212. **R. El-Mehdawi, F. R. Fronczek, and D. M. Roundhill,** *Inorg. Chem.,* 1986, *25*, 3714.
213. **G. Rawji, M. Hediger, and R. M. Milburn,** *Inorg. Chim. Acta,* 1983, *79*, 246.
214. **H. Basch, M. Krauss, W. J. Stevens, and D. Cohen,** *Inorg. Chem.,* 1986, *25*, 684.
215. **J. P. Charland and A. L. Beauchamp,** *Inorg. Chem.,* 1986, *25*, 4870.
216. **H. I. Heitner, S. J. Lippard, and H. R. Sunshine,** *J. Am. Chem. Soc.,* 1972, *94*, 8936.
217. **H. I. Heitner and S. J. Lippard,** *Inorg. Chem.,* 1974, *13*, 815.
218. **J. H. J. den Hartog, H. van den Elst, and J. Reedijk,** *J. Inorg. Biochem.,* 1984, *21*, 83.
219. **N. Hadjiliadis and T. Theophamides,** *Inorg. Chem. Acta,* 1976, *16*, 67.
220. **N. Farrell,** *J. Chem. Soc. Chem. Commun.,* 1980, 1014.
221. **B. Lippert,** *Inorg. Chem.,* 1981, *20*, 4326.
222. **M. Gullotti, C. Pacchioni, A. Pasini, and R. Ugo,** *Inorg. Chem.,* 1982, *21*, 2006.
223. **A. Terzis and D. Meutzafos,** *Inorg. Chem.,* 1983, *22*, 1140.
224. **R. Bau, H. K. Choi, R. C. Stevens, and S. K. Shang Huang,** in *5th Int. Symp. on Platinum and Other Metal Coordination Compounds in Cancer Chemotherapy,* M. Nicolini and G. Bandoli, Eds., Cleup Padua, Italy, 1987, 341.
225. **R. Faggiani, C. J. L. Lock and B. Lippert,** *J. Am. Chem. Soc.,* 1980, *102*, 5418.
226. **B. Lippert,** *Inorg. Chim. Acta,* 1981, *56*, L23.
227. **A. T. M. Marcelis, J. C. van der Veer, J. C. M. Zwetsloot, and J. Reedijk,** *Inorg. Chim. Acta,* 1983, *78*, 195.
228. **J. R. Lusty, H. S. O. Chan, S. Ang, and P. F. Lee,** unpublished work.
229. **M. D. Reily, K. Wilkowski, K. Shinozuka, and L. G. Marzilli,** *Inorg. Chem.,* 1985, *24*, 37.
230. **P. Laverture, J. Hubert, and A. L. Beauchamps,** *Inorg. Chem.,* 1976, *15*, 322.
231. **E. A. Griffith and E. L. Amma,** *J. Chem. Soc. Chem. Commun.,* 1979, 1013.
232. **E. Sletten and A. Apeland,** *Acta Crystallogr. Sect. B,* 1975, *B31*, 2019.
233. **A. L. Shoemaker, P. Singh, and D. J. Hodgson,** *Acta Crystallogr. Sect. B,* 1976, *B32*, 979.
234. **L. Pope, M. Laing, M. R. Caira, and R. Nassimbeni,** *Acta Crystallogr. Sect. B,* 1976, *B32*, 612.
235. **T. J. Kistenmacher, L. G. Marzilli, and D. J. Szalda,** *Acta Crystallogr. Sect. B,* 1976, *B32*, 186.
236. **S. Kirschner, Y.-K. Wei, D. Francis, and J. G. Bergman,** *J. Med Chem.,* 1966, *9*, 369.
237. **B. Rosenberg,** *Biochimie,* 1978, *60*, 859.
238. **C. J. L. Lock, R. A. Speranzini, G. Turner, and J. Powell,** *J. Am. Chem. Soc.,* 1976, *98*, 7865.
239. **A. P. Hitchcock, C. J. L. Lock, W. M. C. Pratt, and B. Lippert,** in *Platinum, Gold and Other Metal Chemotherapeutic Agents* (ACS Symp. 209), American Chemical Society, Washington, D.C., 1983.
240. **B. Pullman and A. Pullman,** *Quantum Biochemistry,* Interscience, New York, 1963.
241. **P. de Meester and A. C. Skapski,** *J. Chem. Soc. Dalton Trans.,* 1973, 1596.
242. **K. Tomita, T. Izumo, and T. Fujiwara,** *Biochem. Biophys. Res. Commun.,* 1973, *54*, 96.
243. **R. Weiss and H. Venner,** *Z. Physiol. Chem.,* 1963, *333*, 169.
244. **P. de Meester and A. C. Shapski,** *J. Chem. Soc. A,* 1971, 2167.

245. **A. Terzis, A. L. Beauchamp, and R. Rivest,** *Inorg. Chem.,* 1973, *12*, 1166.
246. **M. V. Hanson, C. B. Smith, G. D. Simpson, and G. O. Carlisle,** *Inorg. Nucl. Chem. Lett.,* 1975, *11*, 225.
247. **D. L. Kozlowski and D. J. Hodgson,** *J. Chem. Soc. Dalton Trans.,* 1975, 55.
248. **D. J. Szalda, T. J. Kistenmacher, and L. G. Marzilli,** *Inorg. Chem.,* 1975, *14*, 2623.
249. **E. Sletten and B. Thorstensen,** *Acta Cyrstallogr. Sect. B,* 1974, *B30*, 2483.
250. **E. Sletten and M. Rudd,** *Acta Cyrstallogr. Sect. B,* 1975, *B31*, 982.
251. **P. de Meester, D. M. L. Goodgame, A. C. Shapski, and Z. Warnke,** *Biochim. Biophys. Acta,* 1973, *324*, 301.
252. **M. J. McCall and M. R. Taylor,** Acta *Crystallogr Sect. B,* 1976, *B32*, 1687.
253. **M. J. McCall and M. R. Taylor,** *Biochim. Biophys. Acta,* 1975, *390*, 137.
254. **E. Sletten,** *J. Chem. Soc. Chem. Commun.,* 1967, 1119.
255. **E. Sletten,** *Acta Cyrstallogr. Sect. B,* 1969, *B25*, 1480.
256. **T. J. Kistenmacher,** *Acta Crystallogr. Sect. B,* 1974, *B30*, 1610.
257. **T. J. Kistenmacher, L. G. Marzilli, and C. H. Chang,** *J. Am. Chem. Soc.,* 1973, *95*, 5817.
258. **P. de Meester, D. M. L. Goodgame, K. A. Price, and A. C. Shapski,** *J. Chem. Soc. Chem. Commun.,* 1970, 1573.
259. **P. de Meester and A. C. Shapski,** *J. Chem. Soc. Dalton Trans.,* 1972, 2400.
260. **P. de Meester, D. M. L. Goodgame, K. A. Price, and A. C. Shapski,** *Biochim. Biophys. Acta,* 1971, *44*, 510.
261. **H. -J. Korte and R. Bau,** *Inorg. Chim. Acta,* 1983, *79*, 251.
262. **A. S. Dimoglo, Y. M. Chumakov, and I. B. Bersuber,** *Teor. Eksp. Khim.,* 1981, *17*, 88.
263. **R. B. Martin and Y. H. Miriam,** *Met. Ions Biol. Syst.,* 1979, *8*, 55.
264. **K. P. Beaumont, C. A. McAuliffe, and M. E. Friedman,** *Inorg. Chim. Acta,* 1977, *25,* 241.
265. **W. M. Beck, J. C. Calabrese, and M. D. Kottmair,** *Inorg. Chem.,* 1979, *18,* 176.

Section 5

COMPLEXES INVOLVING NUCLEOSIDES OF THE PURINE BASES

Virtudes Moreno, Angel Terrón, Juan Alabart, Kenji Inagaki, and Yoshinori Kidani

INTRODUCTION

The purine nucleosides constitute one of the largest groups of all the transition metal complexes. While most studies, and indeed this introduction, have concentrated on platinum and palladium complexes, many other transition metals have been used in complex formation.

Much of the earlier work using purine bases focused on simple nucleobases derived principally from adenosine and guanosine. However, it soon became apparent that the behavior of these bases did not necessarily duplicate the interaction of transition metals, particularly platinum, with corresponding nucleosides and nucleotides.

The ease of study of some of the first row transition metals, their inexpensive costs, and the range of spectroscopic and analytical techniques available have increased the interest in elements such as cobalt, chromium, nickel, and copper. The reader is therefore referred to publications and reviews of these complexes for further details of their study. The remainder of this introduction concentrates on the platinum and palladium complexes.

Platinum and Palladium Complexes

Since the discovery of antitumor active *cis*-dichlorodiammineplatinum(II), *cis*-$Pt(NH_3)_2Cl_2$, many platinum ammine compounds having *cis*-geometry, i.e., $Pt(LL)Cl_2$, L=monodentate amine and LL=bidentate amine, such as ethylenediamine, have been prepared. Investigation on the mechanism of action of *cis*-$Pt(LL)Cl_2$, though it is still ongoing, indicates that an important biological target of the platinum compounds is cellular DNA.[202] With reference to these investigations, research in the area of platinum compounds has been directed toward platinum-nucleobase interaction. Binding of platinum compounds to protein and RNA, in addition to DNA, has been also known to occur.[202] Investigations on ternary complexes of Pt(II) (or Pd(II)) with nucleosides and amino acids or peptides have been undertaken in order to characterize nucleic acid-metal-protein interactions.[105,138,149,187]

Since Pt(II) is a soft metal ion, platinum compounds are likely to bind preferentially to the nitrogen atoms on purine bases, but binding to the sugar moiety of nucleosides is unlikely to occur. Therefore, in the N9-alkyl-purines, the sugar moiety in nucleosides is replaced by an alkyl group, and these are thought to be good model compounds for exploring platinum coordination sites.[6,203-215] The amine ligands, LL in $Pt(LL)Cl_2$, are inert in a substitution reaction with purine nucleosides. Therefore, the amine ligands bound to metal ions are generally retained in the products without substitution, and the reader can see such a formula in the references.

Palladium amine compounds ($Pd(LL)Cl_2$), which are diamagnetic and square planar complexes, do not show antitumor activity. However, Pd(II) amine compounds form complexes with nucleosides in a similar binding mode as platinum compound. Antitumor inactivity of Pd-amine compounds may be related to a fast ligand exchange reaction. This has been known to be 10^5 times faster compared to Pt(II). Since ligand exchange reactions of Pt(II) are quite slow, studies on the interaction of Pd(II) with nucleosides are expected to be useful as models for Pt(II) nucleosides interaction at equilibrium. Octahedral Pt(IV) amine compounds also have an antitumor activity, but only a limited number of Pt(IV) nucleosides have been reported. It should be noted that reactions of Pt(IV) amine compounds with nucleosides yields Pt(II) nucleosides as a final reaction product.[114,216]

Pt(II) and Pd(II) amine compounds preferentially bind to nitrogen atoms of base moiety, especially N7 of 6-oxopurine derivatives and N7 and N1 of adenine derivatives. Binding of Pt(II) and Pd(II) amine compounds to the N1 site of 6-oxopurine is likely to occur after deprotonation at the N1. The amino groups in purine nucleosides, i.e., $6\text{-}NH_2$ in adenine derivatives and $2\text{-}NH_2$ in guanine derivatives, are very weak metal binding sites because the lone pair of electrons of the $6\text{-}NH_2$ groups delocalize into the π-system of the purine ring. At least, there is no binding of Pt(II) and Pd(II) to the amino group under the neutral and acidic conditions. In alkaline pH, the $6\text{-}NH_2$ group of adenine derivatives becomes a potential metal binding site (after deprotonation of the NH_2 group). This is not the case for the $2\text{-}NH_2$ group of 6-oxopurine. Since deprotonation of the N1 takes place in preference to that of the $2\text{-}NH_2$, the N1 is a much more favorable metal binding site. Inosine derivatives, in which the $2\text{-}NH_2$ group of guanine is replaced by a hydrogen atom, are therefore good model compounds of guanosine.

In the reaction of $Pt(LL)C1_2$ with 6-oxopurine derivatives, the possibility of the existence of a direct metal O6 interactions, in addition to N7 bonding, has been the subject of much debate. Only brief views will be given here. For details of the arguments for and against the N7,O6 chelation, the reader is referred to the original literature.

Views for N7,O6 chelation hypothesis:[48,69,73,146,150,215] at pH >9, 6-oxopurine derivatives give negatively charged oxygen atoms at O6 due to the tautomeric conversion of keto form to enol form, and this leads to the formation of N7,O6 chelate complexes (or polymers with N7,O6 bridge). In the infrared spectra of the resulting compounds, the carbonyl stretching band (C(6)=O) is shifted to lower frequencies by about 75 cm^{-1}, being presented as evidence of a direct platinum (or palladium)-carbonyl interaction.

Views against N7,O6 chelation hypothesis:[96,127,172,182,212,217,218] the above N7,O6 chelate compounds may be assigned to polymers with N1 and N7 bonding without O6 coordination. The pK_a value at the N1 lowers by 1.5 to 2.0 units upon platination at N7. Lowering of the pK_a at the N1 makes this position a better donor atom. This becomes accessible for N1 platination. Additionally, (1) X-ray studies of 6-oxopurine-platinum complexes reported so far do not show any coordination of O6; (2) the data of ^{1}H- and ^{195}Pt-NMR and pH titration do not support a hypothetical structure for N7,O6 chelate with N1 deprotonated; and (3) the shift to lower frequencies of the carbonyl stretching vibration is also observed upon ionization of N1 iminoproton.

Consistent views: the first binding of bifunctional platinum compounds occurs at the N7 of 6-oxopurines. In acidic pH, the reaction of platinum compounds with 6-oxopurines yields only N7-coordinated complexes, i.e., the second binding also occurs at the N7.

The reaction of bifunctional platinum compounds with nucleosides occurs via a two-step mechanism.[219] The monofunctional intermediate, the species formed upon the first binding step, can rotate about the Pt-nucleoside bond (e.g., Pt-N7 bond in the case of guanosine) in search of a second binding site. The presence of bulky substituents on amine ligands, e.g., *N,N,N′,N′*-tetramethylethylenediamine, leads to restricted rotation of nucleosides about the Pt-N7 bond on the NMR time scale.[46,114] When two nucleosides in $Pt(LL)Cl_2$ are in a head-to-tail arrangement, slow rotation of nucleosides about the Pt-N7 bond results in diastereoisomers because of the presences of chiral sugar moieties.

Some excellent reviews have been published and are listed in the references.

TABLE INDEX

Base no.	Compound	Base
A1n	Adenosine	ado
A2n	Adenosine-1-oxide	adolox
A3n	Deazaadenosine (tubercidin)	dazado
A4n	2′-Deoxiadenosine	2′dado
A5n	*N*(6),*N*(6)-Dimethyladenosine	dmeado
A6n	1,*N*(6)-Ethenoadenosine	ε -ado
A7n	*N*(6)-(Δ^2-Isopentenyl)adenosine	ipentado
A8n	1-Methyladenosine	1meado
A9n	*N*(6)-Methyladenosine	6meado
A10n	2′,3′-*O*-Isopropylideneadenosine	isopado
A11n	Tetracetyladenosine	teado
A12n	2′,3′,5′-Triacetyladenosine	traado
G1n	Guanosine	guo
G2n	8-Bromoguanosine	8Brguo
G3n	2′-Deoxiguanosine	2′dguo
G4n	*N*(2),*N*(2)-Dimethylguanosine	dmeguo
G5n	1-Methylguanosine	1meguo
G6n	1-Methyl-2′deoxiguanosine	1me2′guo
G7n	7-Methylguanosine	7meguo
G8n	2′,3′-*O*-Isopropylideneguanosine	isopguo
G9n	6-Thioguanosine	6Sguo
G10n	8-Thioguanosine	8Sguo
G11n	2′,3′,5′-Triacetylguanosine	traguo
I1n	Inosine	ino
I2n	7-Deazainosine	dazino
I3n	2′-Deoxiinosine	2′dino
I4n	1-Methylinosine	1meino
I5n	7-Methylinosine	7meino
I6n	2′,3′,-*O*-Isopropylideneinosine	isopino
I7n	2′,3′,5′-Triacetylinosine	traino
P1n	Purineriboside [9(β-D-ribofuranosyl) purine]	rfpur
P2n	*N*-[9-(β-D-ribofuranosyl)purin-6-ylcarbamoyl]threonine	t6a
P3n	2-Amino-6-mercaptopurineriboside	2A6Spurr
P4n	6-Thiopurineriboside	6Spurr
P5n	6-Methoxypurineriboside	6meopurr
X1n	Xanthosine	xao
X2n	7-Methylxanthosine	7mexao

TABLE 1

Base no.	Base	Metal	Stoichiometry	Method	Ref.
			Adenosine		
Aln	ado	Ag(I)	Ag...ado	pot, uv	95
		Au(III)	*cis*-$[Me_2Au(ado)Cl]$	nmr	128
			$[Au(ado)_3Cl](OH)_2$	nmr, ir, uv	32
			Au...ado	uv, eph, chrom	65
				kin	31
		Cd(II)	$[Cd(ado)]Cl_2$	esr, ir	16
			$[Cd(ado)_2]Cl_2$	esr, ir	16
			$CdCl_2$...ado...DMSO	nmr, ks	176
			Cd...ado	nmr, ks	170
		Cr(III)	$[Cr_2(ado)_3](ClO_4)_6$	ir	126
		Co(II)	$[Co(ado)](ClO_4)_2$	ir	126
			$[Co(ado)_2]Cl_2$	esr, ir mag, spec	16
			Co...ado	nmr, pot	109
				nmr	88
			$CoCl_2$..ado..guo..DMSO	nmr	88
			Co...ado...DMSO	nmr	117
			Co(bipy)...ado	spec, ks	33
		Co(III)	$Co_3(OH)_5(ado)_2(H_2O)_6(NH_2)_2$	cond, pot, ir	17, 18
				uv	41
			$Co_3(OH)_5(ado)_2(H_2O)_5(NH_2)_2$	cond, pot, ir	17, 18
				uv	41
			$Co_3(OH)_4(ado)_2(H_2O)_4(NH_2)_2Cl$	cond, pot, ir	17, 18
				uv	41
			$Co_3(OH)_4(ado)_3(H_2O)_5(NH_2)_2$	cond, pot, ir	17, 18
				uv	41
			$Co_3(OH)_4(ado)_3(H_2O)_4(NH_2)_2$	cond, pot, ir	17, 18
				uv	41
			$[Co(acac)_2(NO_2)(ado)]$	nmr, ks	181
			Co(dien)...ado	uv, CD	184
		Cu(II)	$[Cu(ado)(OH)]$	esr, ir, uv, mag	134
			$[Cu(ado)]\ (ClO_4)_2$	ir	126
			$[Cu(ado)]\ Cl_2$	esr, ir	16
			$[Cu(ado)_2]\ Cl_2$	mag, spec	16
			Cu...ado	nmr, pot	109
				esr, ir, nmr, mag	120
				nmr	14
				uv, ks	97
				nmr, esr	30
				nmr, pot	109
				pot, ks	78

TABLE 1 (CONTINUED)

Base no.	Base	Metal	Stoichiometry	Method	Ref.
				pol	183
				cond, uv, ks	60
				pot, ks	81
			$Cu(NO_3)_2$...ado	uv, pot	52
			Cu(glygly)...ado	esr, uv	50
			Cu(glygly)...ado...DMSO	nmr	117
			Cu(bipy)...ado	spec, ks	33
				uv	130
			Cu...ado...trp	nmr, pot, uv, ks	178
			(Cl)(NMeN′SEN)Cu...ado...DMSO	nmr	117
		Fe(II)	$[Fe(ado)](ClO_4)_2$	ir	126
		Fe(III)	$[Fe_2(ado)_3](ClO_4)_6$	ir	126
		Hg(II)	$HgCl_2$...ado	nmr	19
				nmr, ram, ks	116
			$HgCl_2$...ado...DMSO	nmr, ks	176
				nmr	89
			MeHg...ado	ks	179
				ram	113
				nmr	186
			Hg...ado	uv	51, 193
			Hg...ado...DMSO	nmr	84
			$HgCl_2$...guo...ado...DMSO	nmr	89
			$HgCl_2$...ado...cyd...DMSO	nmr	89
		Ir(III)	$[IrCl_2(H_2O)_3(ado)]Cl$	ir, cond	94
		La(III)	$La(NO_3)_3$...ado	nmr, ram	116
		Mn(II)	$[Mn(ado)](ClO_4)_2$	ir	126
			$(\pi\text{-}MeC_5H_4)Mn(CO)_2(ado)$	ir	11
			Mn...ado	nmr, hch	5
				nmr, esr, ks	80
			Mn...ado...trp	nmr, pot, uv, ks	178
			Mn...ado...DMSO	nmr	117
		Ni(II)	$[Ni(ado)](ClO_4)_2$	ir	126
			$[Ni(ado)]Cl_2$	ir, mag, spec	16
			Ni...ado	nmr, pot	109
				kin	27
				kin, ks, uv	185
				ks, uv	115
			Ni(bipy)...ado	spec, ks	33
			Ni...ado...DMSO	nmr	117
		Os(VI)	$[OsO_2(py)_2(ado)]$	X-ray	42
				kin, uv, nmr, ir, chrom	47
			$[OsO_2(bipy)(ado)]$	kin, uv, nmr, ir, chrom	47
		Pd(II)	$[Pd(ado)_2Cl_2]$	uv, ir	196
				uv	34

TABLE 1 (CONTINUED)

Base no.	Base	Metal	Stoichiometry	Method	Ref.
			$([Pd(NH_3)_2Cl]_2(ado))Cl_2$	nmr, cond, ir	151
			$[Pd(ado)_2Cl_2]$.2DMSO	nmr, cond, uv	54
			$[Pd(ado)_2Br_2]$.2DMSO	uv, nmr, cond	54
			$[Pd(dien)(ado)]^{2+}$	nmr	172
			$[(Pd(dien))_2(ado)]^{4+}$	nmr	172
			$[Pd(ado)Cl_2]$	nmr, ir, cond, uv	54
			$[Pd(ado)Br_2]$	nmr, ir, cond, uv	54
			[Pd(Omecys)(ado)Cl]	nmr, cond, ir	153
			$[(Pd(pro)Cl)_2(\mu\text{-ado})]$	nmr, cond, ir	148
			$[Pd(Bu_3P)Cl_2]_2(\mu\text{-ado})$	nmr, cond	12
				nmr	11
			cis-$[Pd(caf)(ado)Cl_2]$	nmr, ir	150
			$[Pd(ado)_4]Cl_2$	nmr, ir	49
			$[Pd(\eta^3\text{-}C_3H_5)(ado)Cl]$	nmr, ir	168
			$[Pd(\eta^3\text{-}C_3H_5)Cl]_2(ado)$	nmr, ir	168
			$([Pd(pmt)_2(ado)]Cl_2)_2$	nmr, ir, uv, cond	152
			(en)Pd...ado	nmr	77
				nmr, pot	182
			$[Pd(en)(ado)_2]^{2+}$	nmr,	77
			(dien)Pd...ado	nmr, ks	96
				nmr, ks	172
				pot, nmr	182
				uv, ks	97
			$(dien)Pd(H_2O)$...ado	nmr	107
			(pmdien)Pd...ado	nmr, ks	96
			glyasp-Pd...ado	nmr	105
		Pt(II)	$[MePt(COD)(ado)]NO_3$	nmr	98
			$[Pt(dat)(ado)_2]Cl_2$	nmr, ir	48
			cis-$[Pt(gly)(ado)Cl_2)]$	ir, uv	93
			cis-$[Pt(ala)(ado)Cl_2]$	ir	93
			[Pt(Omecys)(ado)Cl]	nmr, ir, cond	74
			$[(Pt(Omecys)Cl)_2(\mu\text{-ado})]$	nmr, cond	74
			$[(Pt(Bu_3P)Cl_2)_2(\mu\text{-ado})]$	nmr, cond	12
			cis-$[Pt(caf)(ado)Cl_2]$	nmr, ir, cond	150
			$[(Pt(acac)Cl)_2(\mu\text{-ado})]$	nmr	55
			$[(PtCl_2(DMSO))_2(\mu\text{-ado})]$	nmr	55
			$[Pt(ado)_4]Cl_2$	nmr, ir	49
			$[Pt(ado)_4](PF_6)_2$	ir	49
			$([Pt(pmt)_2(ado)]Cl_2)_2$	nmr, ir, uv, cond	152
			$K_2[(PtCl_3)_2(\text{ -ado})]$	nmr, cond	101
			trans-$[Pt(ado)_2Cl_2]$	nmr, ir, uv, mag, cond	72
			trans-$[Pt(ado)_2Br_2]$	uv, mag, cond, ir, mag	68

TABLE 1 (CONTINUED)

Base no.	Base	Metal	Stoichiometry	Method	Ref.
			cis-$[Pt(ado)_2Cl_2]$	nmr, cond	70
			cis-$[Pt(ado)_2Br_2]$	nmr, cond	70
			$[Pt(ado)Cl_3]$	nmr, cond	70
			$[Pt(ado)I_2]$	nmr, ir, uv	10
			$[Pt(en)(ado)_2]Cl_2$	nmr	99
			$[Pt(dien)(ado)]^{2+}$	nmr	99
				nmr	107
			$[(Pt(NH_3)_3)_2(\mu\text{-ado})]^{4+}$	nmr	83
			cis-$[PtCl_2(NH_3)_2]$...ado	spec	112
				spec	136
				spec, ks	173
				pot, uv, ks	83
			$[(Pt(dien))_2(\mu\text{-ado})]^{4+}$	nmr	99
				nmr	107
			trans-$[PtCl_2(NH_3)_2]$...ado	spec	112
			cis-$[PtCl_2(DMSO)_2]$...ado	nmr, ram	116
			$[Pt(NH_3)_3Cl]Cl$...ado	pot, uv, ks	83
			$[Pt(en)Cl_2]$...ado	chrom	166
		Pr(III)	$Pr(NO_3)_3$...ado	nmr, ram	116
		Rh(I)	$[Rh(PPh_3)_2(CO)(ado)]PF_6$	nmr	2
			$[Rh(CO)_2(ado)Cl]$	nmr, ir	180
			$[(PPh_3)_2(CO)Rh]$.ado.thd	nmr	1
			$[(PPh_3)_2(CO)Rh]$.ado.cyd	nmr	1
			$[(PPh_3)_2(CO)Rh]$.ado.urd	nmr	1
		Rh(II)	$[Rh_2(ac)_4](ado)$	nmr, ir, uv	154
			$[Rh_2(ac)_4][(ado)]$	nmr	55
				nmr	56
		Rh(III)	$RhCl_3(ado)(MeOH)_2$	ir, cond	94
			$[RhCl(NH_3)_4(ado)]Cl_2$	ir, cond	94
			$[RhCl_3(ado)(DMSO)_2]$	nmr	55
		Ru(III)	$[Ru(NH_3)_5(ado)]Br_3$	uv, ech	38
			$[Ru(NH_3)_5(ado)]^{2+}$	uv, ech	38
			$RuCl_3(ado).(MeOH)_2$	ir, cond	94
			$[Ru(ado)_4(H_2O)_2](ClO_4)_3$	nmr, ir, uv, cond	95
				pot	144
		W(0)	$W(CO)_5(ado)$	nmr, ir	11
		U(VI)	UO_2...ado	pot	59
			UO_2...ado...GLP	pot	59
		Zn(II)	$[Zn(ado)](ClO_4)_2$	ir	126
			$[Zn(ado)_2]Cl_2$	esr, ir	16
			Zn...ado	nmr, pot,	109
				uv, ks	97
				nmr, ram	116
				nmr, titr	109
				nmr, ks	170
			Zn...ado...DMSO	nmr, ks	191

TABLE 1 (CONTINUED)

Base no.	Base	Metal	Stoichiometry	Method	Ref.
			$ZnCl_2$...ado...DMSO	nmr, ks	176
			$ZnCl_2$...ado...guo...DMSO	nmr	88
			Zn...ado...guo...DMSO	nmr	191
			Zn...ado...cyd...DMSO	nmr	191
			$Zn(NO_3)_2$...ado	nmr	19
				nmr, ram	116
			Zn(bipy)...ado	spec, ks	33
			Zn...ado...trp	nmr, pot, spec, ks	178
			Adenosine-1-oxide		
A2n	ado1ox	Co(II)	Co...ado1ox	ks, uv, pot	143
		Cu(II)	Cu...ado1ox	ks, uv, pot	143
		Fe(II)	Fe...ado1ox	ks, uv, pot	143
		Mn(II)	Mn...ado1ox	ks, uv, pot	143
		Ni(II)	Ni...ado1ox	ks, uv, pot	143
		Pt(II)	$([Pt(ado1ox)Cl])_n$	ir, cond	70
			$([Pt(ado1ox)Br])_n$	cond, ir	70
			[(Pt(en)(ado1ox)]Cl	cond	70
			[(Pt(dien)Cl]Cl...ado1ox	nmr	99
		Zn(II)	Zn..ado1ox	ks, uv, pot	143
			7-Deazaadenosine (Tubercidin)		
A3n	dazado	Cu(II)	Cu...dazado	nmr, ir, esr, mag	120
				nmr	14
			(en)Cu...dazado	nmr	57
			Cu...dazino...dazado	nmr	123
		Rh(I)	$[Rh(CO)_2(dazado)Cl]$	nmr, ir	180
		Ru(III)	$[Ru(NH_3)_5(dazado)]^{3+}$	uv, ech	38
			$[Ru(NH_3)_5(dazado)]^{2+}$	uv	38
			2′Deoxiadenosine		
A4n	2′dado	Ag(I)	Ag...2′dado	pot	66
		Co(III)	$[Co(acac)_2NO_2(2'dado)].3.5H_2O$	X-ray, nmr, ks	181
			(dien)Co...2′dado	uv, CD	184
		Cu(II)	Cu...2′dado	nmr, ir, esr, mag	120
				pot, ks	81
		Mn(II)	Mn...2′dado	nmr, hch	5
				nmr	61
		Pd(II)	*trans*-$[Pd(2'dado)_2(SCN)_2]$	nmr, ir	25

TABLE 1 (CONTINUED)

Base no.	Base	Metal	Stoichiometry	Method	Ref.
			***N*(6),*N*(6)-Dimethyladenosine**		
A5n	dmeado	Pt(II)	*cis*-$[Pt(NH_3)_2Cl_2]$...dmeado	spec	112
			trans-$[Pt(NH_3)_2Cl_2]$...dmeado	spec	112
			1,*N*(6)-Ethenoadenosine		
A6n	ε-ado	Cu(II)	Cu…ε-ado	nmr, uv, pot	171
			Cu...bipy...ε-ado	nmr, uv, pot	171
		Mn(II)	Mn…ε-ado	nmr, uv, pot	171
		Zn(II)	Zn...ε-ado	nmr, uv, pot	171
			***N*(6)-(Δ^2-Isopentenyl)adenosine**		
A7n	ipent-ado	Os(VI)	$[OsO_2(py)_2$(ipentado)] Olefinic ester	nmr, ir, uv	155
			$[OsO_2(py)_2$(ipentado)] Sugar ester	nmr, ir, uv	155
			$[OsO_2$(bipy)(ipentado)] Sugar ester	nmr, ir, uv	155
			$[OsO_2$(bipy)(ipentado)] Olefinic ester	nmr, ir, uv ks	155
			1-Methyladenosine		
A8n	1meado	Au(III)	[Au(1meado)Cl_2]	ir	32
		Cu(II)	Cu...1meado	nmr, esr, mag ir	120
			(en)Cu...1meado	nmr	57
		Hg(II)	MeHg...1meado	ram	113
		Pt(II)	*cis*-$[Pt(NH)_2Cl_2]$.1meado	spec, ks	112
				spec	136
			trans-$[Pt(NH_3)_2Cl_2]$.1meado	spec	112
			$[Pt(NH_3)_3Cl]Cl$.1meado	pot, uv, ks	83
		Ru(III)	$[Ru(NH_3)_5(1meado)]^{3+}$	uv, ech	38
			N(6)-Methyladenosine		
A9n	6meado	Pt(II)	[Pt(dien)Cl]Cl...6meado	nmr	99
			2′,3′-*O*-Isopropylideneadenosine		
A10n	isop-ado	Pt(II)	$[Pt(Bu_3P)Cl_2]_2$(isopado)	nmr, cond	12

TABLE 1 (CONTINUED)

Base no.	Base	Metal	Stoichiometry	Method	Ref.
			Tetraacetyladenosine		
A11n	teaado	Pt(II)	[Pt(OMecys)(teaado)Cl]	cond	74
			trans-$[Pt(teaado)_2Cl_2]$	nmr, ir	72
				ir	68
			trans-$[Pt(teaado)_2Br_2]$	ir, mag, cond	72
		Rh(II)	$[Rh_2(ac)_4]$.2(teaado)	nmr, ir, uv	154
			2′3′5′Triacetyladenosine		
A12n	traado	Pt(II)	Pt(OMecys)(traado)Cl	nmr, ir, cond	74
			cis-$[Pt(traado)_2Cl_2]$	nmr, ir, cond	70
			cis-$[Pt(traado)_2Br_2]$	cond	70
		Rh(II)	$[Rh_2(ac)_4]$(traado)	nmr, ir, uv	154
			Guanosine		
G1n	guo	Ag(I)	[Ag(guo)]	nmr	37
			Ag...guo	uv, cond, ir	190
				uv	194
				pot, uv	53
		Au(I)	$Au(guo)_2Cl$	nmr, ir, MB, cond	24
				MB, ir	23
				nmr, ir, cond	75
		Au(III)	$[Au(guo)_2Cl_2]Cl$	MB	23
				nmr, ir, cond	75
			$Au(guo)Cl_3$	nmr, ir, MB, cond	24
				MB, ir	23
				nmr, ir, cond	75
			$Au(guo)Cl_2$	nmr, ir, MB, cond	24
				MB	23
				ir, cond	75
				nmr, ir, uv	32
			$[Au(guo)_4]Cl_2(OH)$	nmr, ir, uv	32
			cis-$[Me_2Au(guo)Cl]$	nmr	128
			Au...guo	kin	31
		Cd(II)	Cd...guo	ks	4
				nmr, ks	170
			$CdCl_2$...guo...DMSO	nmr	195
				nmr, ks	176
			$CdBr_2$...guo...DMSO	nmr	195
			CdI_2...guo...DMSO	nmr	195
		Co(II)	Co...guo	ks	4
				pot, ks	64
				ks, pot	108

TABLE 1 (CONTINUED)

Base no.	Base	Metal	Stoichiometry	Method	Ref.
			Co...guo...DMSO	nmr	88
			Co...bipy...guo	pot, ks	64
			Co...ado...guo...DMSO	nmr	88
		Co(III)	$[Co(acac)_2(NO_2)(guo)]$	ks, nmr	181
		Cu(II)	$Cu(guo)_2$	esr, ir, uv	132
			Cu...guo	uv, cond, pot, ir	189
				uv, ks	97
				nmr, esr	30
				ks	4
				ks, pot	108
				pol	183
				cond, uv, ks	60
				esr, ir	122
				ir	61
			$Cu(NO_3)_2$...guo	uv, pot	52
			Cu(glygly)...guo	esr, uv	50
		Fe(II)	Fe...guo	ks	4
		Hg(II)	[(MeHg)(guo)]	nmr, ir	26
				nmr	20
				nmr, ir	20
			$[(MeHg)(guo)]NO_3$	nmr, ir	26
				nmr, ir	20
			$[(MeHg)_2(guo)]NO_3$	nmr, ir	26
				nmr, ir	20
			(PhHg)(guo)	nmr, ir	26
			$[(PhHg)_2(guo)]NO_3$	nmr, ir	26
			$[(PhHg)(guo)]NO_3$	nmr, ir	26
			[(PhHg)(guo)]	ir	142
			Catena-(μ-Cl)-Cl(guo)Hg	X-ray	7
			Hg...guo	uv	193
				nmr	118
				uv	51
			MeHg...guo	uv, ks	179
				nmr	21
				nmr	186
			$HgCl_2$...guo	nmr	19
			$HgCl_2$...guo...DMSO	nmr	84
				nmr	89
				nmr, ks	176
			$HgCl_2$...guo...ado...DMSO	nmr	89
			$HgCl_2$...guo...cyd...DMSO	nmr	89
		Ir(I)	$[Ir(PPh_3)_3(CO)(guo)]$	ir	11
		La(III)	La...guo	nmr	118
		Lu(III)	Lu...guo	nmr	118
		Mn(II)	Mn..guo	ks	4
		Ni(II)	Ni...guo	uv, ks	97

TABLE 1 (CONTINUED)

Base no.	Base	Metal	Stoichiometry	Method	Ref.
				ks	4
				ks, pot	108
				ks, pot	64
			Ni...dmeado...guo	ks, uv	6
			Ni-bipy...guo	ks, pot	64
		Os(VI)	$[OsO_2(py)_2(guo)]$	kin, nmr, ir, uv, chr	47
			$[OsO_2(bipy)(guo)]$	nmr, ir, uv kin, chr	47
		Pd(II)	$K[Pd(guo)Cl_3]$	ir, cond	73
			$[Pd_2(guo)_2]Cl_2$	cond, ir	73
			cis-$[Pd(guo)_2Cl_2]$	nmr, ir	147
				nmr, ir, cond	146
			$[Pd(guo)_2]Cl_2$	nmr, ir	25
			$[Pd(guo)_4]Cl_4$	nmr, ir, cond	146
			trans-$[Pd(guo)_2]$	cond	146
			trans-$[Pd(guo)_2Cl_2$	cond	146
			cis-$[Pd(guo)_2(H_2O)_2]Cl_2$	nmr, ir, cond	146
				nmr, ir	147
			cis-$[Pd(guo)_2]$	nmr, ir, cond	146
				nmr, ir	147
			cis-$[Pd(guo)_2(guo)]Cl$	nmr, ir, cond	146
				nmr, ir	147
			$[Pd(guo)_3Cl]Cl$	nmr, ir, cond	146
				nmr, ir	147
			trans-$[Pd(guo)_2(tu)_2]Cl_2$	nmr	146
			Pd(acac)(guo)Cl	nmr	55
			$[Pd(NH_3)_2(guo)_2]Cl_2$	nmr, ir, cond	151
			$[Pd(NH_3)_2(DMSO)(guo)]Cl_2$	nmr, ir, cond	151
			$Pd(NH_3)_2(guo)Cl$	cond	151
			$[Pd(NH_3)_2(guo)Cl]Cl$	cond	151
			$[Pd(dien)(guo)](ClO_4)_2$	X-ray	167
			$[Pd(dien)(guo)]Cl_2$	nmr	167
			$[Pd(dmen)(guo)_2]$	nmr	131
			$[Pd(guo)_3(xao)]Cl_2$	nmr, ir, cond	73
			$[Pd(guo)(ino)_3]Cl_2$	cond, nmr, ir	73
			$[Pd(guo)_3(ino)]Cl_2$	cond, nmr, ir	73
			trans-$[Pd(guo)(ino)Cl_2]$	ir, cond	73
			cis-$[(Pd(guo)(xao)Cl_2]$	cond, ir	73
			trans-$[Pd(guo)(ino)]$	cond, ir	73
			cis-$[Pd(guo)_2(ino)_2]Cl_2$	nmr, ir, cond	146
			trans-$[Pd(guo)_2(ino)_2]Cl_2$	nmr, ir, cond	146
			cis-$[Pd(guo)_2(ino)]Cl$	nmr, ir, cond	146
			$[Pd(guo)_2(ino)Cl]Cl$	nmr, ir, cond	146
			cis-$[Pd(guo)_2(cyd)_2]Cl_2$	ir, cond	146
			trans-$[Pd(guo)_2(cyd)_2]Cl_2$	ir, cond	146
			cis-$[Pd(guo)_2(xao)_2]Cl_2$	ir, cond	146

TABLE 1 (CONTINUED)

Base no.	Base	Metal	Stoichiometry	Method	Ref.
			trans-[(Pd(guo)$_2$(xao)$_2$]Cl_2	ir, cond	146
			cis-[Pd(gly)(guo)$_2$]Cl	nmr, ir, cond	90
			cis-[Pd(gly)(guo)$_2$Cl]Cl	nmr, ir, cond	90
			cis-[Pd(ile)(guo)$_2$]Cl	nmr, ir, cond	90
			cis-[Pd(ile)(guo)$_2$Cl]Cl	nmr, ir, cond	90
			cis-[Pd(val)(guo)$_2$]Cl	nmr, ir, cond	90
			cis-[Pd(val)(guo)$_2$Cl]Cl	nmr, ir, cond	90
			cis-[Pd(pro)(guo)$_2$]Cl	nmr, ir, cond	90
			cis-[Pd(pro)(guo)$_2$Cl]Cl	nmr, ir, cond	90
			cis-[Pd(ala)(guo)$_2$]Cl	nmr, ir, cond	90
			cis-[Pd(ala)(guo)$_2$Cl]Cl	nmr, ir, cond	90
			cis-[Pd(phe)(guo)$_2$]Cl	nmr, ir, cond	90
			cis-[Pd(phe)(guo)$_2$Cl]Cl	nmr, ir, cond	90
			[Pd(Omecys)(guo)Cl]	nmr, ir, cond	153
			[Pd(Omecys)(guo)]	nmr, ir, cond	153
			[Pd(pro)(guo)Cl]	nmr, ir, cond	148
			[Pd(pro)(guo)]	cond	148
			cis-[Pd(caf)(guo)Cl_2]	nmr, ir, cond	150
			Pd(caf)(guo)Cl	cond, nmr, ir	150
			[Pd(mit)$_2$(guo)$_2$]Cl_2	nmr, ir	49
			Pd(η^3-C_3H_5)(guo)Cl	nmr, ir	168
			[(Bu_3P)Pd(guo)Cl]$_n$	cond	12
			(Bu_3P)$_2$Pd(guo)$_2$	cond	12
				nmr, ir	11
			cis-[Pd(pmt)$_2$(guo)$_2$]Cl_2	nmr, ir, cond, uv	152
			Pd(en)...(guo)$_2$	nmr	77
		Pt(II)	[Pt(guo)$_2$$I_2$]	nmr, ir, uv	10
			[Pt(guo)$_2$$Cl_2$].2HCl	nmr, ir	69
			Pt(guo)$_2$$Cl_2$	nmr, cond	69
			[Pt(guo)$_4$]Cl_2	nmr, cond	69
			[Pt(guo)$_2$(ino)$_2$]Cl_2	nmr, cond	69
			Pt(guo)$_2$	ir, cond	69
			K[Pt(guo)Cl_3]	nmr, cond	101
			[Pt(NH_3)$_2$(guo)(OH_2)]$^{2+}$	nmr, uv, HPLC	198
				nmr	199
			cis-[Pt(NH_3)$_2$(guo)Cl]Cl	uv	141
				nmr, titr, chr	127
			cis-[Pt(NH_3)$_2$(guo)$_2$]$Cl_{3/2}$(ClO_4)$_{1/2}$ O.7H_2	nmr, uv, CD, atta, X-ray	43
				X-ray	44
			cis-[(Pt(NH_3)$_2$(guo)$_2$](ClO_4)$_2$	nmr	58
			cis-[(Pt(NH_3)$_2$(guo)$_2$]Cl_2	uv	141
				atta	43
				nmr, titr, chr	127
				CD	67
			cis-[(Pt(NH_3)$_2$(guo)$_2$]Cl_2	nmr, ir, uv	102

TABLE 1 (CONTINUED)

Base no.	Base	Metal	Stoichiometry	Method	Ref.
			cis-$[Pt(NH_3)_2(guo)_2]^{2+}$	nmr, ir	137
			$[Pt(NH_3)_3(guo)]^{2+}$	nmr	137
			cis-$[(Pt(NH_3)_2Cl)_2\mu\text{-guo}]^+$	nmr, titr, chr	127
			cis-$[(Pt(MeNH_2)_2Cl)_2(\mu\text{-guo})]^+$	nmr, chr	137
				nmr,	127
			cis-$[(Pt(MeNH_2)_2(guo)_2]^{2+}$	nmr, chr	127
			cis-$[Pt(MeNH_2)_2(guo)Cl]^+$	nmr, chr	127
			$[Pt(en)(guo)_2]Cl_{3/2}I_{1/2}.2H_2O$	X-ray	9, 63
			$[Pt(en)(guo)_2]Cl_2$	CD	139
			$[Pt(en)(guo)_2]Cl_2.2Me_2CO$	CD, uv, cond	67
			$[Pt(en)(guo)_2]Cl_2$	nmr	102
			$[Pt(en)(guo)_2]^{2+}$	nmr, chr	127
				nmr, ir	137
			[Pt(en)(guo)]Cl	nmr, ir, cond, uv, CD	139
			$[Pt(en)(guo)]NO_3$	cond, uv, CD, nmr, ir	139
			$[Pt(en)(guo)]ClO_4$	nmr, ir, cond	139
			$[Pt(en)(guo)](NO_3)_2$	CD, uv, cond	67
				nmr, ir, cond, uv, CD	139
				CD	138
			$[Pt(en)(guo)Cl]^+$	nmr, chr	127
			$[(Pt(en)Cl)_2(\ \text{-guo})]^+$	nmr	127
			$[Pt(R,R\text{-dmen})(guo)_2]^{2+}$	nmr	45
			$[Pt(S,S\text{-dmen})(guo)_2]^{2+}$	nmr	45
			$[Pt(R,S\text{-dmen})(guo)_2]^{2+}$	nmr	45
			$[Pt(tmen)(guo)_2]^{2+}$	nmr	46
			cis-$[Pt_2(en)_2(guoCl_2)]^+$	nmr, chr	127
			$[(Pt(dien)(guo)](ClO_4)_2$	X-ray	125
			$[Pt(dien)(guo)]^{2+}$	nmr	137
				nmr, CD	197
			cis-$[Pt(val)(tba)(guo)_2]Cl_2$	nmr, ir, CD	138
			cis-$[Pt(leu)(tba)(guo)_2]Cl_2$	nmr, ir	138
			cis-$[Pt(gly)(tba)(guo)_2]Cl_2$	ir, CD, uv	138
			cis-$[Pt(phe)(tba)(guo)_2]Cl_2$	ir, CD	138
			cis-$[Pt(pro)(tba)(guo)_2]Cl_2$	ir, CD	138
			cis-$[Pt(ser)(tba)(guo)_2]Cl_2$	CD	138
			cis-$[Pt(ala)(tba)(guo)_2]Cl_2$	ir	138
			$[Pt(SMecys)(guo)_2]Cl_2$	nmr, ir, cond	187
			$[Pt(SEtcys)(guo)_2]Cl_2$	nmr, ir, cond	187
			$[Pt(Sbzcys)(guo)_2]Cl_2$	ir, cond	187
			$[Pt(5.NOBzcys)(guo)_2]Cl_2$	cond, ir	187
			[Pt(OMecys)(guo)Cl]	nmr, ir, cond	74

TABLE 1 (CONTINUED)

Base no.	Base	Metal	Stoichiometry	Method	Ref.
			cis-[Pt(gly)(guo)$_2$]Cl	nmr, ir, cond	90
			cis-[Pt(gly)(guo)$_2$Cl]Cl	nmr, ir, cond	90
			cis-[Pt(ile)(guo)$_2$]Cl	nmr, ir, cond	90
			cis-[Pt(ile)(guo)$_2$Cl]Cl	nmr, ir, cond	90
			cis-[Pt(val)(guo)$_2$]Cl	nmr, ir, cond	90
			cis-[Pt(val)(guo)$_2$Cl]Cl	nmr, ir, cond	90
			cis-[Pt(pro)(guo)$_2$]Cl	nmr, ir, cond	90
			cis-[Pt(pro)(guo)$_2$Cl]Cl	nmr, ir, cond	90
			cis-[Pt(ala)(guo)$_2$]Cl	nmr, ir, cond	90
			cis-[Pt(ala)(guo)$_2$Cl]Cl	nmr, ir, cond	90
			cis-[Pt(phe)(guo)$_2$]Cl	nmr, ir, cond	90
			cis-[Pt(phe)(guo)$_2$Cl]Cl	nmr, ir, cond	90
			cis-[Pt(gly)(guo)Cl$_2$]	nmr, ir, uv	93
			cis-[Pt(ala)(guo)Cl$_2$]	ir	93
			[Pt((*R*)-pn)(guo)$_2$]Cl$_2$.Me$_2$CO	nmr, CD, uv, cond	67
			[Pt((*S*)-pn)(guo)$_2$]Cl$_2$.Me$_2$CO	cond, nmr, CD, uv	67
			[Pt((*S*,*S*)-bn)(guo)$_2$]Cl$_2$.Me$_2$CO	nmr, CD, uv, cond	67
			[Pt((*R*,*R*)-dach)(guo)$_2$]Cl$_2$	nmr, uv cond	67
			[Pt((*S*,*S*)-dach)(guo)$_2$]Cl$_2$	nmr, CD, uv, cond	67
			[Pt((*R*)-pen)(guo)$_2$]Cl$_2$	nmr, CD, uv, cond	67
			[Pt((*R*,*R*)-bn)(guo)$_2$]Cl$_2$.0.5Me$_2$CO	CD, cond, nmr	67
			[Pt((meso)-bn)(guo)$_2$]Cl$_2$.1.5Me$_2$CO	CD, cond, nmr	67
			[Pt((meso)-dach)(guo)$_2$]Cl$_2$	CD, cond, nmr	67
			[Pt(*R*,*S*-dach)(guo)$_2$]$^{2+}$	X-ray	200
			[Pt((*R*,*R*)-dpen)(guo)$_2$]Cl$_2$	CD, cond	67
			Pt((*S*,*S*)-dpen)(guo)$_2$]Cl$_2$	CD, cond	67
			[Pt((meso)-dpen)(guo)$_2$]Cl$_2$	CD,cond	67
			[Pt((*R*,*R*)-dach)(guo)](NO$_3$)$_2$	CD, cond	67
			[Pt((*S*,*S*)-dach)(guo)](NO$_3$)$_2$	CD, cond	67
			[Pt(bn)(guo)]Cl	nmr, ir, cond, uv, CD	139
			[Pt(bn)(guo)]I	uv, CD, nmr, ir, cond	139
			cis-[Pt(caf)(guo)Cl$_2$]	nmr, ir, cond	150
			Pt(caf)(guo)Cl	nmr, ir, cond	150
			[Pt(mit)$_2$(guo)$_2$]Cl$_2$	nmr, ir	49
			[Pt(mit)$_2$(guo)$_2$](PF$_6$)$_2$		49
			[Pt(opda)(guo)$_2$]Cl$_2$	nmr, ir, uv, cond	106
			[Pt(dmopda)(guo)$_2$]Cl$_2$	nmr, ir, uv, cond	106
			[Pt(bipy)(guo)$_2$]Cl$_2$	nmr, ir, uv, cond	106
				nmr	114

TABLE 1 (CONTINUED)

Base no.	Base	Metal	Stoichiometry	Method	Ref.
			$[Pt(bpe)(guo)_2](NO_3)_2$	nmr	114
			cis-$[Pt(py)_2(guo)_2]Cl_2$	nmr	102
			cis-$[Pt(py)_2(guo)_2]Cl_2.H_2O$	nmr	114
			trans-$[Pt(py)_2(guo)_2]Cl_2$	nmr	102
			cis-$[Pt(\alpha\text{-pic})_2(guo)_2]Cl_2$	nmr	114
			cis-$[Pt(\alpha\text{-pic})_2(guo)_2](NO_3)_2$	nmr	114
			cis-$[Pt(pmt)_2(guo)_2]Cl_2$	nmr, ir, uv, cond	152
			$[MePt(COD)(guo)]NO_3$	nmr	98
			$[Pt(dat)(guo)]Cl_2$	nmr, ir	48
			$[Pt(dat)(guo)Cl]Cl$	nmr, ir	48
			$[Pt(dmdap)(guo)_2]Cl_2$	nmr	114
			$[Pt(tmdap)(guo)_2]Cl_2$	nmr	114
			$[Pt(dmdap)(guo)_2](NO_3)_2$	nmr	114
			$[Pt(tmdap)(guo)_2](NO_3)_2$	nmr	114
			$[Pt(tmtn)(guo)_2]Cl_2$	nmr	114
			$[Pt(dmtn)(guo)_2]Cl_2$	nmr	114
			$[Pt(dmtn)(guo)_2](NO_3)_2$	nmr	114
			$[Pt(acac)(guo)Cl]$	nmr	55
			$[Pt(DIPSO)(guo)Cl_2]$	nmr	55
			$[Pt(dad)(guo)_2]^{2+}$	nmr	137
			$[Pt(date)(guo)_2]^{2+}$	nmr	137
			$[Pt(datr)(guo)_2]^{2+}$	nmr, ir	137
			$[Pt(DMSO)(guo)Cl_2]$	nmr	103
				nmr	55
			$Pt(DMSO)_2$...guo	nmr	118
			cis-$[Pt(NH_3)_2]$...guo	spec	112
				spec, ks	173
			trans-$[Pt(NH_3)_2Cl_2]$...guo	spec	112
			$[Pt(en)Cl_3]$...guo	chr	166
				nmr	86
			Pt(dien)...guo	nmr, kin	135
		Pr(III)	Pr...guo	nmr	118
		Rh(I)	$[Rh(PPh_3)_2(CO)(guo)]PF_6$	nmr, ir	2
			$[Rh(PPh_3)_2(CO)(guo)]$	ir	11
			$[Rh(CO)_2(guo)Cl]$	nmr, ir	180
			$[Rh(CO)_2(guo)_2]Cl$	nmr, ir, cond	149
			$[Rh(guo)_3Cl]$	nmr, ir, cond	149
			$[(PPh_3)_2(CO)Rh]^+$...guo	nmr	3
			$[(PPh_3)_2(CO)Rh]^+$...guo...tea	nmr	3
			$PPh_3)_2(CO)Rh]^+$...guo...EOA	nmr	3
			$[(PPh_3)_2(CO)Rh]^+$...guo...ado	nmr	1
			$[(PPh_3)_2(CO)Rh]^+$...guo...cyt	nmr	1
		Rh(III)	$[Rh(CO)_2(guo)_2Cl_2]Cl$	nmr, ir, cond	149
			$Rh(guo)_3Cl_3$	nmr, ir, cond	149
		Ru(III)	$[Ru(NH_3)_5(guo)]Cl_3$	uv, ech, kin	39
			$[Ru(NH_3)_5(guo)]Cl_3$	ir, uv, cond	92
			$[Ru(guo)_5(H_2O)](ClO_4)_3$	nmr, ir, uv, cond	95

TABLE 1 (CONTINUED)

Base no.	Base	Metal	Stoichiometry	Method	Ref.
		Zn(II)	Zn...guo	uv, ks	97
				nmr	118
				ks, pot	108
				ks	4
				ks, nmr	170
			Zn...guo...DMSO	nmr, ks	191
			Zn...guo...ado...DMSO	nmr	191
				nmr	88
			$Zn(NO_3)_2$...guo...ado	nmr	19
			$ZnBr_2$...guo...DMSO	nmr	195
			$ZnCl_2$...guo...DMSO	nmr, ks	176
			ZnI_2...guo...DMSO	nmr	195
			Zn...imz...guo...DMSO	nmr	88
			8-Bromoguanosine		
G2n	8Brguo	Cu(II)	$[Cu(8Brguo)_2]$	esr, ir, uv	132
			2′-Deoxyguanosine		
G3n	2′dguo	Cu(II)	$[Cu_3(2'dguo)_2(OH)_4]$	esr, ir, uv, mag	133
			Cu...2′dguo	esr	122
				pot, ks	81
				ir	61
		Hg(II)	Hg...2′dguo	nmr	118
		Mn(II)	Mn...2′dguo	nmr, hch	5
		Pt(II)	*cis*-$[Pt(NH_3)_2Cl_2]$...2′dguo	uv	156
			$[Pt(en)Cl_2]$...2′guo	chr	167
		Ru(III)	$[Ru(NH_3)_5(2'dguo)]^{3+}$	uv, ech	38
			***N*(2),*N*(2)-Dimethylguanosine**		
G4n	dme-guo	Hg(II)	Hg...dmeguo	nmr	118
		Pr(III)	Pr...dmeguo	nmr	118
		Pt(II)	$[Pt(NH_3)_2(dmeguo)_2]^{2+}$	nmr, uv, HPLC	198
			$[Pt(NH_3)_2(dmeguo)(OH_2)]^{2+}$	nmr, uv, HPLC	198
			1-Methylguanosine		
G5n	1me-guo	Cu(II)	Cu...1meguo	nmr, ir	122
		Hg(II)	Hg...1meguo	nmr	118
			(MeHg)...1meguo	nmr	186
		Pd(II)	$[Pd(en)(1meguo)_2]^{2+}$	nmr	131
		Pr(III)	Pr...1meguo	nmr	118
		Pt(II)	*cis*-$[Pt(NH_3)_2(1meguo)Cl]^+$	nmr, chr	127
			cis-$[Pt(NH_3)_2(1meguo)_2]^{2+}$	nmr, chr	127
			cis-$[Pt(NH_3)_2Cl_2]$...1meguo	spec	112
				spec, ks	136

TABLE 1 (CONTINUED)

Base no.	Base	Metal	Stoichiometry	Method	Ref.
			trans-$[Pt(NH_3)_2Cl_2]$...1meguo	spec	112
		Rh(I)	$Rh(CO)_2$(1meguo)Cl	nmr	180
		Ru(III)	$[Ru(NH_3)_5$(1meguo)]Cl_3	uv, kin, ech	39
			$[Ru(NH_3)_4$(1meguo)Cl]Cl_2	kin, ech	39
		Zn(II)	Zn...1meguo	nmr	118
			1-Methyl-2′-Deoxyguanosine		
G6n	1me2′-dguo	Ru(III)	$[Ru(NH_3)_5$(1me2′dguo)Cl_3	uv, ech, kin	39
			7-Methylguanosine		
G7n	7me-guo	Au(III)	Au(7meguo)Cl_2	ir	32
		Cu(II)	Cu...7meguo	nmr, ir	122
		Hg(II)	Hg...7meguo	uv, ks	179
			MeHg...7meguo	uv, ks	179
		Pr(III)	Pr...7meguo	nmr	177
		Pt(II)	*cis*-$[Pt(NH_3)_2Cl_2]$.7meguo	spec	136
			[Pt(en)Cl_2]...7meguo	chr	166
		Zn(II)	Zn...7meguo	nmr	177
			2′,3′-*O*-Isopropylideneguanosine		
G8	isop-guo	Rh(I)	*cis*-$[Rh(CO)_2$(isopguo)Cl]	nmr, ir	11
		W(0)	$[W(CO)_5$(isopguo)]	nmr, ir	104
			6-Thioguanosine		
G9	6Sguo	Hg(II)	$HgCl_2$...6Sguo...DMSO	nmr	79, 84
			(MeHg)...6Sguo...DMSO	nmr	84
			8-Thioguanosine		
G10	8Sguo	Hg(II)	$HgCl_2$...8Sguo...DMSO	nmr	79, 84
			(MeHg)—8Sguo...DMSO	nmr	84
			2′3′5′-Triacetylguanosine		
G11	traguo	Au(I)	$[Au(traguo)_2Cl]$	nmr, ir, MB, cond	24
		Au(III)	[Au(traguo)Cl_3]	MB	23
				ir, cond	75
			[Au(traguo)Cl_2]	nmr, ir, MB, cond	24
				ir, cond	75
				MB	23
			$[Au(traguo)_2Cl_2]Cl$	MB	23
				nmr, ir, cond	75

TABLE 1 (CONTINUED)

Base no.	Base	Metal	Stoichiometry	Method	Ref.
		Pd(II)	$K[Pd(traguo)Cl_3]$	ir, cond	73
		Pt(II)	[Pt(Omecys)(traguo)Cl]	ir, cond, nmr	74
			Inosine		
I1n	ino	Ag(I)	Ag...ino	cond, uv, ir	190
		Au(I)	$[Au(ino)_2Cl]$	MB, ir	23
				nmr, ir, MB, cond	24
				nmr, ir, cond	75
			$[Au(PPh_3)(ino)]$	nmr, ir	168
			Au...ino	uv	58
		Au(III)	$[Au(ino)Cl_3]$	ir, MB	23
				nmr, ir, cond	75
			$[Au(ino)Cl_2]$	MB	23
				nmr, ir, MB, cond	24
				ir	32
				ir, cond	75
			$[Au(ino)_2Cl_2]Cl$	MB	23
				nmr, ir, cond	75
		Cd(II)	Cd...ino	ks	4
				nmr, ks	170
		Co(II)	Co...ino	ks	4
				ks, pot	108
				ir	62
			(bipy)Co...ino	spec, ks	33
		Co(III)	$[Co(en)_2(ino)Cl]Cl_2$	nmr, ir, uv,	129
			(dien)Co...ino	uv, CD	184
		Cu(II)	[Cu(ino)(OH)]	esr, ir, uv, mag	134
			[Cu(ino)]	esr, mag	119
			$[Cu(ino)_2O]_n$	mag, esr	119
			Cu...ino	cond. pot, uv ir	189
				nmr, uv	15
				nmr, esr, uv, mag, CD	119
				ks	4
				ks, pot	108
				uv, ks	97
				nmr, ir, esr	123
				nmr, ir	61
				nmr	62
			(bipy)Cu...ino	spec, ks	33
				uv	130
		Fe(II)	Fe...ino	ks	4

TABLE 1 (CONTINUED)

Base no.	Base	Metal	Stoichiometry	Method	Ref.
		Hg(II)	$[(MeHg)(ino)]NO_3$	nmr, ir	20
			$[(MeHg)_2(ino)]NO_3$	nmr, ir	20
			[(MeHg)(ino)]	nmr, ir	20
			$[(MeHg)_2(\mu\text{-ino})]ClO_4$	X-ray	13
			$[(MeHg)_3(ino)]NO_3$	nmr	21
				nmr, ir	20
			[(PhHg)(ino)]	ir	142
			(MeHg)...ino	uv, ks	179
				nmr, ram, ir	110
				nmr, ram	111
				nmr	21
			Hg...ino	uv	51
				nmr	118
				uv, ks	179
		Ir(I)	$[Ir(PPh_3)_2(CO)(ino)]$	ir	11
		Mn(II)	Mn...ino	ks	4
				ir	62
		Ni(II)	Hg...ino	uv, ks	97
				ks	4
				ks, pot	108
				kin	27
			(bipy)Hg...ino	spec, ks	33
		Pd(II)	$K[Pd(ino)Cl_3]$	ir, cond	73
			$[Pd_2(ino)_2Cl_2]$	ir, cond	73
			cis-$[Pd(ino)(xao)Cl_2]$	ir, cond	73
			trans-$[Pd(ino)(guo)Cl_2]$	ir, cond	73
			trans-[Pd(guo)(ino)]	ir, cond	73
			$[Pd(guo)(ino)_3]Cl_2$	nmr, ir, cond	73
			$[Pd(guo)_3(ino)]Cl_2$	nmr, ir, cond	73
			$[Pd(ino)_3(xao)]Cl_2$	nmr, ir, cond	73
			cis-$[Pd(ino)_2Cl_2]$	nmr, ir, cond	146
			cis-$[Pd(ino)_2(xao)_2]Cl_2$	nmr, ir, cond	146
			$[Pd(ino)_4]Cl_2$	nmr, ir, cond	146
			cis-$[Pd(guo)_2(ino)_2]Cl_2$	nmr, ir, cond	146
			trans-$[Pd(guo)_2(ino)_2]Cl_2$	nmr, ir, cond	146
			cis-$[Pd(guo)_2(ino)]Cl$	nmr, ir, cond	146
			cis-$[Pd(guo)_2(ino)Cl]Cl$	nmr, ir, cond	146
			cis-$[Pd(ino)_2]$	ir, cond, nmr	146
			trans-$[Pd(ino)_2]$	ir, cond, nmr	146
			trans-$[Pd(ino)_2Cl_2]$	ir, cond, nmr	146
			trans-$[Pd(ino)_2(xao)_2]Cl_2$	ir, cond, nmr	146
			cis-$[Pd(cyd)_2(ino)_2]Cl_2$	ir, cond, nmr	146
			trans-$[Pd(cyd)_2(ino)_2]Cl_2$	ir, cond, nmr	146
			trans-$[(Pd(ino)_2(tu)_2]Cl_2$	nmr	146
			$[Pd(NH_3)_2(ino)_2]Cl_2$	nmr, cond, ir	151
			$[Pd(NH_3)_2(ino)(DMSO)]Cl_2$	cond, ir, nmr	157
			$[Pd(NH_3)_2(ino)]Cl$	ir, cond	151

TABLE 1 (CONTINUED)

Base no.	Base	Metal	Stoichiometry	Method	Ref.
			$[Pd(NH_3)_2(ino)Cl]Cl$	ir, cond	151
			$[Pd(dien)(ino)]^{2+}$	nmr	167, 172, 182
			$[(Pd(dien))_2(\mu\text{-ino})]^{3+}$	nmr	172, 182
			cis-$[Pd(caf)(ino)Cl_2]$	nmr, ir, cond	150
			[Pd(caf)(ino)]Cl	nmr, ir, cond	150
			$[Pd(\eta^3\text{-}C_3H_5)(ino)Cl]$	nmr, ir	168
			cis-$[Pd(pmt)_2(ino)_2]Cl_2$	nmr, ir, uv, cond	152
			[Pd(Omecys)(ino)Cl]	nmr, ir, cond	153
			[Pd(Omecys)(ino)]	nmr, ir, cond	153
			[Pd(pro)(ino)Cl]	nmr, ir, cond	148
			[Pd(pro)(ino)]	ir, cond	148
			(en)Pd...ino	nmr	77
				nmr, pot	182
			$(en)PdCl_2$...ino	sfk	87
			(dien)Pd...ino	ks, nmr	96,172
				nmr, ks	172
				nmr, pot	182
				uv, ks	97
			$[Pd(dien)Br]^+$...ino	sfk	175
			(pmdien)Pd...ino	nmr, ks	96
		Pt(II)	$[Pt(opda)(ino)_2]Cl_2$	nmr, ir, uv, cond	106
			$[Pt(dmopda)(ino)_2]Cl_2$	nmr, ir, cond, uv	106
			$[Pt(bipy)(ino)_2]Cl_2$	nmr, uv, ir, cond	106
			cis-$[Pt(NH_3)_2(ino)_2]Cl_2.Me_2CO$	nmr	102
			$[Pt(en)(ino)_2]Cl_2.Me_2CO$	nmr	102
			cis-$[Pt(NH_3)_2(ino)_2](ClO_4)_2$	nmr	58
			$[Pt(dien)(ino)](NO_3)_2.H_2O$	X-ray	124
			$[Pt(dien)(ino)]Cl_2$	nmr	167
			$[Pt(en)(ino)_2]^{2+}$	nmr, cond	69
				ram, nmr	36
				ram, nmr	188
			cis-$[Pt(DMSO)(ino)Cl_2]$	nmr	103
			trans-$[Pt(DMSO)(ino)Cl_2]$	nmr	103
			cis-$[Pt(ino)_2Cl_2]$	nmr, ir, cond	69
				nmr	100
			cis-$[Pt(ino)_2Br_2]$	nmr	69
			$[Pt(ino)_2Cl]$	nmr	69
			$[Pt(ino)Cl]_n$	ir, cond	69
			$[Pt(ino)_2]$	ir, cond	69
			$[Pt(guo)_2(ino)_2]Cl_2$	nmr, cond	69
			$[Pt(cyd)_2(ino)_2]Cl_2$	cond, nmr	69
			cis-$[Pt(NH_3)_2(ino)_2]Cl_2$	atta	43

TABLE 1 (CONTINUED)

Base no.	Base	Metal	Stoichiometry	Method	Ref.
			cis-$[Pt(pmt)_2(ino)_2]Cl_2$	nmr, ir, uv, cond	152
			$K[(Pt(ino)Cl_3]$	nmr	100, 101
				cond	103
				nmr	100
			$[Pt(ino)_3Cl]Cl$	nmr	100
			$[Pt(ino)_4]Cl_2$	nmr, ir, cond	69
				nmr	100
			$[Pt(Smecys)(ino)_2]Cl_2$	nmr, ir, cond	187
			$[Pt(Setcys)(ino)_2]Cl_2$	nmr, ir, cond	187
			[Pt(Omecys)(ino)Cl]	nmr, ir, cond	74
			cis-$[Pt(gly)(ino)_2]Cl$	nmr, ir, cond	90
			cis-$[Pt(ile)(ino)_2]Cl$	nmr, ir, cond	90
			cis-$[Pt(val)(ino)_2]Cl$	nmr, ir, cond	90
			cis-$[Pt(pro)(ino)_2]Cl$	nmr, ir, cond	90
			cis-$[Pt(ala)(ino)_2]Cl$	nmr, ir, cond	90
			cis-$[Pt(phe)(ino)_2]Cl$	nmr, ir, cond	90
			cis-$[Pt(gly)(ino)_2Cl]Cl$	nmr, ir, cond	90
			cis-$[Pt(ile)(ino)_2Cl]Cl$	nmr, ir, cond	90
			cis-$[Pt(val)(ino)_2Cl]Cl$	nmr, ir, cond	90
			cis-$[Pt(pro)(ino)_2Cl]Cl$	nmr, ir, cond	90
			cis-$[Pt(ala)(ino)_2Cl]Cl$	nmr, ir, cond	90
			cis-$[Pt(phe)(ino)_2Cl]Cl$	nmr, ir, cond	90
			cis-$[Pt(gly)(ino)Cl_2]$	nmr, ir, uv	93
			cis-$[Pt(ala)(ino)Cl_2]$	ir, uv	93
			trans-$[(Bu_3P)_2Pt(ino)_2]$	nmr, ir, cond	12
			$[(Bu_3P)Pt(ino)Cl]_n$	nmr, ir, cond	12
			cis-$[Pt(caf)(ino)Cl_2]$	nmr, ir, cond	150
			[Pt(caf)(ino)]Cl	nmr, ir, cond	150
			cis-$[Pt(NH_3)_2Cl_2]$...ino	nmr, ram	35, 36, 188
				spec	112
			trans-$[Pt(NH_3)_2Cl_2]$...ino	spec	112
			(en)Pt...ino	ram	188
				nmr, ram	36
			(dien)Pt...ino	nmr, kin	135
			[Pt(dien)Br]...ino	kin	174
			$[Pt(dien)(H_2O)]$...ino	kin	174
			$[Pt(dien)(ino)]^{2+}$	kin	174
		Rh(I)	$[Rh(PPh_3)_2(CO)(ino)]PF_6$	nmr, ir	2
			$[Rh(PPh_3)_2(CO)(ino)]$	ir	11
			$[Rh(CO)_2(ino)Cl]$	nmr, ir	180
			$[Rh(CO)_2(ino)_2]Cl$	nmr, ir, cond	149
			$[Rh(ino)_3Cl]$	nmr, ir, cond	149
			$[(PPh_3)_2(CO)Rh]$...ino	nmr	3
			$[(PPh_3)_2(CO)Rh]$...ino...tea	nmr	3
			$[(PPh_3)_2(CO)Rh]$...ino...EOA	nmr	3

TABLE 1 (CONTINUED)

Base no.	Base	Metal	Stoichiometry	Method	Ref.
		Rh(III)	$[Rh(CO)_2(ino)_2Cl_2]Cl$	nmr, ir, cond	149
			$[Rh(ino)_3Cl_3]$	nmr, ir, cond	149
		Ru(II)	$[Ru(NH_3)_5(ino)]^{2+}$	uv, cv	40
			$[Ru(NH_3)_5(ino)]^{+}$	uv, cv	40
		Ru(III)	$[Ru(NH_3)_5(ino)]Cl_3$	cond, uv, ir	92
			$[Ru(NH_3)_5(ino)]Cl_3$	uv, cv	40
			$[Ru(ino)_5(H_2O)](ClO_4)_3$	ir, uv, cond	95
			$[Ru(NH_3)_5(ino)]^{3+}$	uv, ech	38
			$[Ru(NH_3)_5(ino)]^{2+}$	uv	38—40
		Zn(II)	Zn...ino	uv, ks	97
				ks	4
				ks, pot	108
				nmr, ks	170
			(bipy)Zn...ino	spec, ks	33
			7-Deazainosine		
I2n	dazino	Cu(II)	Cu...dazino...dazado	nmr	123
			2′-Deoxyinosine		
I3n	2′dino	Co(III)	(dien)Co...2′dino	uv, CD	184
		Cu(II)	Cu...2′dino	nmr, uv	14
				esr	123
		Mn(II)	Mn...2′dino	nmr, hch	5
			1-Methylinosine		
I4n	1meino	Co(II)	Co...1meino	ks, pot	108
		Cu(II)	Cu...1meino	ks	97
				nmr	57
				ks, pot	108
				nmr, uv	15
				nmr, ir	123
		Hg(II)	(MeHg)...1meino	nmr, ram	111
		Mn(II)	Mn...1meino	nmr	57
		Ni(II)	Ni...1meino	uv, ks	97
				ks, pot	108
		Pd(II)	(dien)Pd...1meino	ks, nmr	95
		Pt(II)	*cis*-$[Pt(NH_3)_2Cl_2]$...1meino	spec	112
				ram, nmr	36
				ram	35
			trans-$[Pt(NH_3)_2Cl_2]$...1meino	spec	112
		Rh(I)	$[Rh(PPh_3)_2(CO)(1meino)]PF_6$	nmr, ir	2
		Ru(II)	$[Ru(NH_3)_5(1meino)]^{2+}$	uv, cv	40
		Ru(III)	$[Ru(NH_3)_5(1meino)]Cl_3$	uv, cv	40
		Zn(II)	Zn...1meino	uv, ks	97
				ks, pot	108

TABLE 1 (CONTINUED)

Base no.	Base	Metal	Stoichiometry	Method	Ref.
			7-Methylinosine		
I5n	7meino	Cd(II)	Cd...7meino	nmr	177
		Co(II)	Co...7meino	ks, pot	108
		Cu(II)	Cu...7meino	uv, ks	97
				ks, pot	108
				nmr, uv	15
		Au(III)	[Au(7meino)Cl_2]	ir	32
		La(III)	La...7meino	nmr	177
		Ni(II)	Ni...7meino	ks, pot	108
		Pt(II)	*trans*-[Pt(DMSO)(7meino)Cl_2]	nmr, cond	165
			cis-[Pt(DMSO)$(7meino)_2$Cl]$^+$	nmr, cond	165
			cis-[Pt$(DMSO)_2$(7meino)Cl]$^+$	nmr, cond	165
			cis-[Pt(7meino)(DMSO)Cl_2]	nmr	177
				nmr, cond	165
			[Pt$(NH_3)_2Cl_2$]...7meino	spec	112
		Pr(III)	Pr...7meino	nmr	177
		Zn(II)	Zn...7meino	uv, ks	97
				nmr	177
				ks, pot	108
			2′,3′-*O*-Isopropilideneinosine		
I6n	isop-ino	Co(III)	(dien)Co...isopino	uv, CD	184
		Cu(II)	Cu...isopino	uv, CD	119
		Mn(II)	Mn...isopino	nmr	57
			2′3′5′-Triacetylinosine		
I7n	traino	Au(I)	[Au$(traino)_2$Cl]	nmr, ir, MB, cond	24
		Au(III)	Au(traino)Cl_3	nmr, ir, MB, cond	24
				MB	23
				ir, cond	75
			[Au(traino)Cl_2]	nmr, ir, MB, cond	24
				ir, cond	75
			[Au$(traino)_2Cl_2$]Cl	MB	23
				nmr, ir, cond	75
		Cu(II)	Cu...traino	nmr, uv	15
		Pd(II)	K[Pd(traino)Cl_3]	ir, cond	73
		Pt(II)	Pt(Omecys)(traino)Cl	nmr, ir, cond	74
			9-(β-D-Ribofuranosyl)purine		
P1n	rfpur	Co(II)	Co...rfpur	nmr, pot	109
		Cu(II)	Cu...rfpur	nmr, pot	109

TABLE 1 (CONTINUED)

Base no.	Base	Metal	Stoichiometry	Method	Ref.
		Ni(II)	Ni...rfpur	nmr, pot	109
		Pt(II)	$([Pt(dien)Cl]Cl)_3$...rfpur	nmr	99
		Zn(II)	Zn...rfpur	nmr, pot	109
			***N*-[9-(β-D-Ribofuranosyl)purin-6-ylcarbamoyl]threonine**		
P2n	t6A	Mn(II)	Mn...t6A	ks, pot	159
			2-Amino-6-mercaptopurineriboside		
P3n	2A6S-purr	Pd(II)	$Pd(2A6Spurr)_2$	EXAFS	201
		Pt(II)	$Pt(2A6Spurr)_2$	nmr, cond	71
				EXAFS	201
			$Pt(2A6Spurr)Cl_2$	cond	71
			$Pt(2A6Spurr)Br_2$	cond	71
			6-Thiopurineriboside		
P4n	6Spurr	Pd(II)	$Pd(6Spurr)_2$	EXAFS	201
		Pt(II)	[Pt(6Spurr)(DMSO)Cl]	nmr, ir, cond	71
			$[Pt(6Spurr)_2]$	nmr, ir, cond	71
				EXAFS	201
			$[Pt(6Spurr)Cl_2]$	nmr, ir, cond	71
			[Pt(en)(6Spurr)]Cl	nmr, cond	71
			$[Pt(6Spurr)_2]$.HCl	nmr, cond	71
			$[Pt(6Spurr)_2]$.2HCl	nmr, cond	71
			$([Pt(6Spurr)Cl])_n$	cond	71
			$[Pt(6Spurr)Br_2]$	cond	71
		Rh(III)	$[Rh(6Spurr)Cl_3]$	nmr, ir, uv, cond, mag	91
		Ru(III)	$[Ru(6Spurr)Cl_3]$	ir, uv, cond, mag	91
			6-Methoxypurineriboside		
P5n	6meo-purr	Cu(II)	Cu-6meopurr	nmr	123
			Xanthosine		
X1n	xao	Cd(II)	Cd...xao	ks	4
		Co(II)	Co...xao	pot, ks	64, 157, 158, 160, 162
				pot	161
				ks	4

TABLE 1 (CONTINUED)

Base no.	Base	Metal	Stoichiometry	Method	Ref.
			Co...xao...tmen	pot, ks	157, 164
			Co...xao...bipy	pot, ks	157, 163, 164
				ks, pot	64
			Co...xao...phen	pot, ks	157, 164
			Co...xao...SSA	pot, ks	157, 164
			Co...xao...his	pot, ks	163
			Co...xao...catechol	pot, ks	163
			Co...xao...oxalic acid	pot, ks	163
			Co...xao...gly	pot, ks	158
		Cu(II)	[Cu(xao)(OH)]	esr, ir, uv, mag	134
			Cu...xao	pot, ks	157, 158, 160, 162
				pot	161
				nmr, ir	121
				ks	4
			Cu...xao...tmen	pot, ks	157, 164
			Cu...xao...bipy	pot, ks	157, 163, 164
			Cu...xao...phen	pot, ks	157, 164
			Cu...xao...SSA	pot, ks	157, 164
			Cu...xao...his	pot, ks	163
			Cu...xao...catechol	pot, ks	163
			Cu...xao...oxalic acid	pot, ks	163
			Cu...xao...gly	pot, ks	158
		Fe(II)	Fe...xao	ks	4
		Hg(II)	[MeHg(xao)]	nmr, ir	22
			$[(MeHg)_2(xao)]$	nmr, ir	22
			$[(MeHg)_3(xao)]NO_3$	nmr, ir	22
			$[MeHg(xao)]NO_3$	nmr, ir	22
			$[(MeHg)_4(xao)]NO_3$	nmr, ir	22
		Ir(III)	$IrCl_3(xao)(MeOH)_2$	cond, ir	94
		Mn(II)	Mn...xao	pot, ks	157, 158 160, 162
				pot	161
				ks	4

TABLE 1 (CONTINUED)

Base no.	Base	Metal	Stoichiometry	Method	Ref.
			Mn...xao...tmen	pot, ks	157, 164
			Mn...xao...bipy	pot, ks	163, 164
			Mn...xao...phen	pot, ks	157, 164
			Mn...xao...SSA	pot, ks	157, 164
			Mn...xao...his	pot, ks	163
			Mn...xao...catechol	pot, ks	163
			Mn...xao...oxalic acid	pot, ks	163
			Mn...xao...gly	pot, ks	158
		Ni(II)	Ni...xao	pot, ks	64, 157, 158, 160, 162
				pot	161
				ks	4
			Ni...xao...tmen	pot, ks	157, 164
			Ni...xao...bipy	pot, ks	64, 157, 163, 164
			Ni...xao...phen	pot, ks	157, 164
			Ni...xao...SSA	pot, ks	157, 164
			Ni...xao...his	pot, ks	163
			Ni...xao...catechol	pot, ks	163
			Ni...xao...oxalic acid	pot, ks	163
			Ni...xao...gly	pot, ks	158
		Pd(II)	$K[Pd(xao)Cl_3]$	nmr, ir, cond	73
			cis-$[Pd(ino)(xao)Cl_2]$	nmr, ir, cond	73
			$[Pd(guo)_3(xao)]Cl_2$	nmr, ir, cond	73
			cis-$[Pd(guo)(xao)Cl_2]$	nmr, ir, cond	73
			$[Pd(ino)_3(xao)]Cl_2$	nmr, ir, cond	73
			$[Pd(dien)(xao)]Cl_2$	nmr	167
			cis-$[Pd(ino)_2(xao)_2]Cl_2$	nmr, ir, cond	146
			trans-$[Pd(ino)_2(xao)_2]Cl_2$	nmr, ir, cond	146
			cis-$[Pd(guo)_2(xao)_2]Cl_2$	nmr, ir, cond	146
			trans-$[Pd(guo)_2(xao)_2]Cl_2$	nmr, ir, cond	146
		Pt(II)	*cis*-$[Pt(NH_3)_2(xao)_2]Cl_2$	nmr	102
			cis-$[Pt(en)(xao)_2]Cl_2$	nmr	102
			cis-$[Pt(NH_3)_2(xao)_2]Cl_2.Me_2CO$	nmr	102
			$[Pt(dien)(xao)]Cl_2$	nmr	167
			$K[Pt(xao)Cl_3]$	nmr, cond	101
			$[Pt(Omecys)(xao)Cl]^+$	nmr, cond, ir	74

TABLE 1 (CONTINUED)

Base no.	Base	Metal	Stoichiometry	Method	Ref.
			cis-$[Pt(NH_3)_2(xao)_2]Cl_2$	atta	43
			cis-$[Pt(DMSO)(xao)Cl_2]$	nmr	103
			trans-$[Pt(DMSO)(xao)Cl_2]$	nmr	103
		Rh(III)	$[Rh(H_2O)_3(xao)Cl_2]Cl$	cond, ir	94
			$[Rh(NH_3)_2(xao)_2Cl_2Cl$	cond, ir	4
		Ru(III)	$[Ru(NH_3)_5(xao)]Cl_3$	uv, cond, ir	92
			$[Ru(H_2O)_5(xao)]Cl_3$	cond, ir	94
		Zn(II)	Zn...xao	pot, ks	157, 158, 160, 162
				pot	161
				ks	4
			Zn...xao...tmen	pot, ks	157, 164
			Zn...xao...bipy	pot, ks	157, 163, 164
			Zn...xao...phen	pot, ks	157, 164
			Zn...xao...SSA	pot, ks	157, 164
			Zn...xao...his	pot, ks	163
			Zn...xao...catechol	pot, ks	163
			Zn...xao...oxalic acid	pot, ks	163
			Zn...xao...gly	pot, ks	158
			7-Methylxanthosine		
X2n	7me-xao	Cu(II)	Cu...7mexao	nmr	121

REFERENCES

1. **D. W. Abbott and C. Woods,** *Inorg. Chem.,* 1983, *22,* 2918.
2. **D. W. Abbott and C. Woods,** *Inorg. Chem.,* 1983, *22,* 597.
3. **D. W. Abbott and C. Woods,** *Inorg. Chem.,* 1983, *22,* 1918.
4. **A. Albert,** *Biochem. J.,* 1953, *54,* 646.
5. **J. A. Anderson, G. P. P. Kuntz, H. H. Evans, and T. J. Swift,** *Biochemistry,* 1971, *10,* 4368.
6. **J. Arpalahti and H. Lönnberg,** *Inorg. Chim. Acta,* 1985, *107,* 197.
7. **M. Authier-Martin, J. Hubert, R. Rivest, and A. L. Beauchamp,** *Acta Crystallogr. Sect. B,* 1978, *B34,* 273.
8. **J. Bariyanga and T. Theophanides,** *Inorg. Chim. Acta,* 1985, *108,* 133.
9. **R. Bau, R. W. Gellert, S. M. Lehovec, and S. Louie,** *J. Clin. Hematol. Oncol.,* 1977, *7,* 51.
10. **K. P. Beaumont and C. A. McAuliffe,** *Inorg. Chim. Acta,* 1977, *25,* 241.
11. **W. Beck and N. Kottmair,** *Chem. Ber.,* 1976, *109,* 970.
12. **W. Beck, J. C. Calabrese, and N. D. Kottmair,** *Inorg. Chem.,* 1979, *18,* 176.
13. **F. Bélanger-Gariépy and A. L. Beauchamp,** *Cryst. Struct. Coommun.,* 1982, *11,* 991.
14. **N. A. Berger and G. L. Eichhorn,** *Biochemistry,* 1971, *10,* 1847.
15. **N. A. Berger and G. L. Eichhorn,** *J. Am. Chem. Soc.,* 1971, *93,* 7062.
16. **T. Beringhelli, M. Freni, F. Morazzoni, P. Romhi, and R. Servida,** *Spectrochim. Acta Part A,* 1981, *3A,* 763.
17. **J. Brigando, and D. Colaïtis,** *Bull. Soc. Chim. Fr.,* 1969, 3449.
18. **J. Brigando, and D. Colaïtis,** *C. R. Acad. Sci., Ser. C,* 1967, *256C,* 867.
19. **G. W. Buchanan and J. B. Stothers,** *Can. J. Chem.,* 1982, *60,* 787.
20. **E. Buncel, A. R., Norris, W. J., Racz, and S. E. Taylor,** *Inorg. Chem.,* 1981, *20,* 98.
21. **E. Buncel. A. R. Norris, W. J., Racz, and S. E. Taylor,** *J. Chem. Soc. Chem. Commun.,* 1979, 562.
22. **E. Buncel, B. K. Hunter, R. Kumar, and A. R. Norris,** *J. Inorg. Biochem.,* 1984, *20,* 171.
23. **G. H. M. Calis and N. Hadjiliadis,** *Inorg. Chim. Acta,* 1984, *91,* 203.
24. **G. H. M. Calis and N. Hadjiliadis,** *Inorg. Chim Acta,* 1983, *79,* 241.
25. **D. Camboli, J. Besançon, J. Tirouflet, B. Gautheron, and P. Meunier,** *Inorg. Chim. Acta,* 1983, *78,* L51.
26. **A. J. Canty and R. S. Tobias,** *Inorg. Chem.,* 1979, *18,* 413.
27. **A. Casper and G. V. Fazakerley,** *J. Chem. Soc. Dalton Trans.,* 1975, 1977.
28. **C. J. Cardin and A. Roy,** *Inorg. Chim. Acta,* 1985, *107,* 57.
29. **C. Chang and L. G. Marzilli,** *J. Am. Chem. Soc.,* 1974, *96,* 3656.
30. **Y. H. Chao and D. R. Kearns,** *J. Am. Chem. Soc.,* 1977, *99,* 6425.
31. **D. Chatterji and S. K. Podder,** *Indian J. Chem.,* 1979, *17A,* 456.
32. **D. Chatterji, U. S. Nandi, and S. K. Podder,** *Biopolymers,* 1977, *16,* 1863.
33. **P. Chaudhuri and H. Sigel,** *J. Am. Chem. Soc.,* 1977, *99,* 3142.
34. **E. G. Chauser, I. P. Rudakova, and A. M. Yurkevich,** *Z. Obshch. Khim.,* 1976, *46,* 360.
35. **G. Y. H. Chu, S. Mansy, R. E. Duncan, and R. S. Tobias,** *J. Am. Chem. Soc.,* 1978, *100,* 593.
36. **G. Y. H. Chu and R. S. Tobias,** *J. Am. Chem. Soc.,* 1976, *98,* 2641.
37. **R. Cini, P. Colamarino, and P. L. Orioli,** *Bioinorg. Chem.,* 1977, *7,* 345.
38. **M. J. Clarke,** *J. Am. Chem. Soc.,* 1978, *100,* 5068.
39. **M. J. Clarke and H. Taube,** *J. Am. Chem. Soc.,* 1974, *96,* 5433.
40. **M. J. Clarke,** *Inorg. Chem.,* 1977, *16,* 738.
41. **D. Colaïtis and J. Brigando,** *Bull. Soc. Chim. Fr.,* 1969, 3453.
42. **J. F. Conn, J. J. Kim, F. L. Suddath, P. Blattmann, and A. Rich,** *J. Am. Chem. Soc.,* 1974, *96,* 7152.
43. **R. E. Cramer, P. L. Dahlstrom, M. J. T. Seu, T. Norton, and M. Kashiwagi,** *Inorg. Chem.,* 1980, *19,* 148.
44. **R. E. Cramer and P. L. Dahlstrom,** *J. Clin. Hematol. Oncol.,* 1977, *7,* 330.

45. **R. E. Cramer and P. L. Dahlstrom,** *Inorg. Chem.,* 1985, *24,* 3420.
46. **R. E. Cramer and P. L. Dahlstrom,** *J. Am. Chem. Soc.,* 1979, *101,* 3679.
47. **F. B. Daniel and E. J. Behrman,** *J. Am. Chem. Soc.,* 1975, *97,* 7352.
48. **J. Dehand and J. Jordanov,** *J. Chem. Soc. Chem. Commun.,* 1976, 598.
49. **J. Dehand and J. Jordanov,** *J. Chem. Soc. Dalton Trans.,* 1977, 1588.
50. **S. V. Deshpande, R. K. Sharma, and T. S. Srivastava,** *Inorg. Chim. Acta,* 1983, *78,* 13.
51. **G. L. Eichhorn and P. Clark,** *J. Am. Chem. Soc.,* 1963, *85,* 4020.
52. **G. L. Eichhorn, P. Clark, and E. D. Becker,** *Biochemistry,* 1966, *5,* 245.
53. **G. L. Eichhorn, J. J. Butzow, P. Clark, and E. Tarien,** *Biopolymers,* 1967, *5,* 283.
54. **R. Ettorre,** *Inorg. Chim. Acta,* 1977, *25,* L9.
55. **N. Farrell,** *J. Chem. Soc. Chem. Commun.,* 1980, 1014.
56. **N. Farrell,** *J. Inorg. Biochem.,* 1981, *14,* 261.
57. **G. V. Fazakerley, G. E. Jackson, M. A. Phillips, and J. C. van Niekerk,** *Inorg. Chim. Acta,* 1979, *35,* 151.
58. **G. V. Fazakerley and K. R. Koch,** *Inorg. Chim. Acta,* 1979, *36,* 13.
59. **I. Feldman, J. Jones, and R. Cross,** *J. Am. Chem. Soc.,* 1967, *89,* 49.
60. **A. M. Fiskin and M. Beer,** *Biochemistry,* 1965, *4,* 1289.
61. **H. Fritzsche and D. Tresselt,** *Stud. Biophys.,* 1970, *24,* 299.
62. **H. Fritzsche,** *Biochim. Biophys. Acta,* 1970, *224,* 608.
63. **R. W. Gellert and R. Bau,** *J. Am. Chem. Soc.,* 1975, *97,* 7379.
64. **R. Ghose and A. K. Dey,** *Recl. Chim. Miner.,* 1980, *17,* 492.
65. **D. W. Gibson, M. Beer, and R. J. Barrnett,** *Biochemistry,* 1971, *10,* 3669.
66. **K. Gillen, R. Jensen, and N. Davidson,** *J. Am. Chem. Soc.,* 1964, *86,* 2792.
67. **M. Gullotti, G. Pacchioni, A. Pasini, and R. Ugo,** *Inorg. Chem.,* 1982, *21,* 2006.
68. **N. Hadjiliadis and T. Theophanides,** *Can. J. Spectrosc.,* 1971, *16,* 135.
69. **N. Hadjiliadis and T. Theophanides,** *Inorg. Chim. Acta,* 1976, *16,* 77.
70. **N. Hadjiliadis and T. Theophanides,** *Inorg. Chim. Acta,* 1976, *16,* 67.
71. **N. Hadjiliadis and T. Theophanides,** *Inorg. Chim. Acta,* 1975, *15,* 167.
72. **N. Hadjiliadis, P. Kourounakis, and T. Theophanides,** *Inorg. Chim. Acta,* 1973, *7,* 226.
73. **N. Hadjiliadis and G. Pneumatikakis,** *J. Chem. Soc. Dalton Trans.,* 1978, 1691.
74. **N. Hadjiliadis and .G. Pneumatikakis,** *Inorg. Chim. Acta,* 1980, *46,* 255.
75. **N. Hadjiliadis, G. Pneumatikakis, and R. Basosi,** *J. Inorg. Biochem.,* 1981, *14,* 115.
76. **A. Hampton and J. C. Fratantoni,** *J. Med. Chem. Acta,* 1966, *9,* 976.
77. **U. K. Häring and R. B. Martin,** *Inorg. Chim. Acta,* 1983, *80,* 1.
78. **T. R. Harking and H. Freiser,** *J. Am. Chem. Soc.,* 1958, *80,* 1132.
79. **H. I. Heitner, S. J. Lippard, and H. R. Sunshine,** *J. Am. Chem. Soc.,* 1972, *94,* 8936.
80. **M. J. Heller, A. J. Jones, and A. T. Tu,** *Biochemistry,* 1970, *9,* 4981.
81. **R. von Hilmer and R. Weiss,** *Z. Phys. Chem.,* 1969, *350,* 1321.
82. **K. Inagaki and Y. Kidani,** *Inorg. Chim. Acta,* 1983, *80,* 171.
83. **K. Inagaki, M. Kumayama, and Y. Kidani,** *J. Inorg. Biochem.,* 1982, *16,* 59.
84. **K. W. Jennette, S. J. Lippard, and D. A. Ucko,** *Biochim. Biophys. Acta,* 1975, *402,* 403.
85. **F. Jordan and B. Y. McFarquhar,** *J. Am. Chem. Soc.,* 1972, *94,* 6557.
86. **J. Jordanov, and R. J. P. Williams,** *Bioinorg. Chem.,* 1978, *8,* 77.
87. **W. Kadima and M. Zador,** *Inorg. Chim. Acta,* 1983, *78,* 77.
88. **L. S. Kan and N. C. Li,** *J. Am. Chem. Soc.,* 1970, *92,* 281.
89. **L. S. Kan and N. C. Li,** *J. Am. Chem. Soc.,* 1970, *92,* 4823.
90. **S. Kasselouri, A. Garoufis, and N. Hadjiliadis,** *Inorg. Chim. Acta,* 1987, *135,* L23.
91. **N. Katsaros and A. Grigoratou,** *J. Inorg. Biochem.,* 1985, *25,* 131.
92. **B. T. Khan, A. Gaffuri, and M. R. Somayajulu,** *Indian J. Chem.,* 1981, *20A,* 189.
93. **B. T. Khan, G. N. Goud, and S. V. Kumari,** *Inorg. Chim. Acta,* 1983, *80,* 145.
94. **B. T. Khan, M. R. Somayajulu, and M. M. T. Khan,** *J. Inorg. Nucl. Chem.,* 1978, *40,* 1251.
95. **B. T. Khan, A. Gaffuri, P. N. Rao, and S. M. Zakeeruddin,** *Polyhedron,* 1987, *6,* 387.
96. **S. Kim and R. B. Martin,** *Inorg. Chim. Acta,* 1984, *91,* 11.

97. **S. Kim and R. B. Martin,** *Inorg. Chim. Acta,* 1984, *91,* 19.
98. **S. Komiya, Y. Mizuno, and T. Shibuya,** *Chem. Lett.,* 1986, 1065.
99. **P. C. Kong and T. Theophanides,** *Inorg. Chem.,* 1974, *13,* 1981.
100. **P. C. Kong and F. D. Rochon,** *Inorg. Chim. Acta,* 1980, *46,* L1.
101. **P. C. Kong and F. D. Rochon,** *J. Chem. Soc. Chem. Commun.,* 1975, 599.
102. **P. C. Kong and T. Theophanides,** *Inorg. Chem.,* 1974, *13,* 1167.
103. **P. C. Kong, D. Iyamuremye, and F. D. Rochon,** *Bioinorg. Chem.,* 1976, *6,* 83.
104. **N. Kottmair and W. Beck,** *Inorg. Chim. Acta,* 1979, *34,* 137.
105. **H. Kozlowski,** *Inorg. Chim. Acta,* 1977, *24,* 215.
106. **L. Kumar and T. S. Srivastava,** *Inorg. Chim. Acta,* 1983, *80,* 47.
107. **M. C. Lim and R. B. Martin,** *J. Inorg. Nucl. Chem.,* 1976, *38,* 1915.
108. **H. Lönnberg and P. Vihanto,** *Inorg. Chim. Acta,* 1981, *56,* 157.
109. **H. Lönnberg and J. Arpalahti,** *Inorg. Chim. Acta,* 1980, *55,* 39.
110. **S. Mansy and R. S. Tobias,** *J. Chem. Soc. Chem. Commun.,* 1974, 957.
111. **S. Mansy and R. S. Tobias,** *Biochemistry,* 1975, *14,* 2952.
112. **S. Mansy, B. Rosenberg, and A. J. Thomson,** *J. Am. Chem. Soc.,* 1973, *95,* 1633.
113. **S. Mansy and R. S. Tobias,** *J. Am. Chem. Soc.,* 1974, *96,* 6874.
114. **A. T. M., Marcelis, J. L. van der Veer, J. C. M. Zwetsloot, and J. Reedijk,** *Inorg. Chim. Acta,* 1983, *78,* 195.
115. **Y. H. Mariam and R. B. Martin,** *Inorg. Chim. Acta,* 1979, *35,* 23.
116. **L. G. Marzilli, B. de Castro, J. P. Caradonna, R. C. Stewart, and C. P. van Vuuren,** *J. Am. Chem. Soc.,* 1980, *102,* 916.
117. **L. G. Marzilli, W. C. Trogler, D. P. Hollis, T. J. Kistenmacher, and C. Chang,** *Inorg. Chem.,* 1975, *14,* 2568.
118. **L. G. Marzilli, B. de Castro, and C. Solorzano,** *J. Am. Chem. Soc.,* 1982, *104,* 461.
119. **K. Maskos,** *Inorg. Chem.,* 1985, *25,* 1.
120. **K. Maskos,** *Acta Biochim. Pol.,* 1978, *25,* 311.
121. **K. Maskos,** *Acta Biochim. Pol.,* 1978, *25,* 303.
122. **K. Maskos,** *Acta Biochim. Pol.,* 1978, *25,* 101.
123. **K. Maskos,** *Acta Biochim. Pol.,* 1978, *25,* 113.
124. **R. Melanson and F. D. Rochon,** *Acta Crystallogr. Sect. B,* 1978, *B34,* 3594.
125. **R. Melanson and F. D. Rochon,** *Can. J. Chem.,* 1979, *57,* 57.
126. **C. M. Mikulski, R. Minutella, N. Defranco, and N. M. Karayannis,** *Inorg. Chim. Acta,* 1985, *106,* L33.
127. **S. K. Miller and L. G. Marzilli,** *Inorg. Chem.,* 1985, *24,* 2421.
128. **D. Mulet, A. M. Calafat, J. J. Fiol, A. Terron, and V. Moreno,** *Inorg. Chim. Acta,* 1987, *138,* 199.
129. **Y. Mizuno and S. Komiya,** *Inorg. Chim. Acta,* 1986, *125,* L13.
130. **C. F. Naumann and H. Sigel,** *J. Am. Chem. Soc.,* 1974, *96,* 2750.
131. **D. J. Nelson, P. L. Yeagle, T. L. Miller, and R. B. Martin,** *Bioinorg. Chem.,* 1976, *5,* 353.
132. **H. C. Nelson and J. F. Villa,** *J. Inorg. Nucl. Chem.,* 1980, *42,* 133.
133. **H. C. Nelson and J. F. Villa,** *Inorg. Chem.,* 1979, *18,* 1725.
134. **H. C. Nelson and J. F. Villa,** *J. Inorg. Nucl. Chem.,* 1979, *42,* 1643.
135. **B. Noszal, V. Scheller-Krattiger, and R. B. Martin,** *J. Am. Chem. Soc.,* 1982, *104,* 1078.
136. **T. O'Connor and W. M. Scovell,** *Chem. Biol. Interact.,* 1979, *26,* 227.
137. **K. Okamoto, V. Behnam, J. Y. Gauthier, S. Hanessian, and T. Theophanides,** *Inorg. Chim. Acta,* 1986, *123,* L1.
138. **A. Pasini and E. Bersanetti,** *Inorg. Chim. Acta,* 1985, *107,* 259.
139. **A. Pasini and R. Mena,** *Inorg. Chim. Acta,* 1981, *56,* L17.
140. **L. Pellerito, G. Ruisi, M. T. Lo Giudice, J. D. Donaldson, and S. M. Grimes,** *Inorg. Chim. Acta,* 1982, *58,* 21.
141. **H. J. Peresie and A. D. Kelman,** *Inorg. Chim. Acta,* 1978, *29,* L247.
142. **P. Peringer,** *Z. Naturforsch. Teil B,* 1979, *34B,* 1459.

143. **D. D. Perrin,** *J. Am. Chem. Soc.,* 1960, *82,* 5642.
144. **R. Phillips and P. George,** *Biochim. Biophys. Acta,* 1968, *162,* 73.
145. **A. C. Plaush and R. R. Sharp,** *J. Am. Chem. Soc.,* 1976, *98,* 7973.
146. **G. Pneumatikakis, N. Hadjiliadis, and T. Theophanides,** *Inorg. Chem.,* 1978, *17,* 915.
147. **G. Pneumatikakis and N. Hadjiliadis,** *Inorg. Chim. Acta,* 1977, *22,* L1.
148. **G. Pneumatikakis,** *Polyhedron,* 1984, *3,* 9.
149. **G. Pneumatikakis, J. Markopoulos, and A. Yannopoulos,** *Inorg. Chim. Acta,* 1987, *136,* L25.
150. **G. Pneumatikakis,** *Inorg. Chim. Acta,* 1984, *93,* 5.
151. **G. Pneumatikakis,** *Inorg. Chim. Acta,* 1982, *66,* 131.
152. **G. Pneumatikakis,** *Inorg. Chim. Acta,* 1982, *46,* 243.
153. **G. Pneumatikakis,** *Inorg. Chim. Acta,* 1983, *80,* 89.
154. **G. Pneumatikakis and N. Hadjiliadis,** *J. Chem. Soc. Dalton Trans.,* 1979, 597.
155. **J. A. Ragazzo and E. J. Behrman,** *Bioinorg. Chem.,* 1976, *5,* 343.
156. **R. O. Rahn, S. S. Chang, and J. D. Hoeschele,** *J. Inorg. Biochem.,* 1983, *18,* 279.
157. **P. R. Reddy, K. V. Reddy, and M. M. T. Khan,** *J. Inorg. Nucl. Chem.,* 1979, *41,* 423.
158. **P. R. Reddy and M. H. Reddy,** *Polyhedron,* 1983, *2,* 1171.
159. **P. R. Reddy, M. P. Schweizer, and G. B. Chheda,** *FEBS Lett.,* 1979, *106,* 63.
160. **P. R. Reddy and K. V. Reddy,** *Inorg. Chim. Acta,* 1983, *80,* 95.
161. **P. R. Reddy K. V. Reddy, and M. M. T. Khan,** *J. Inorg. Nucl. Chem.,* 1976, *38,* 1923.
162. **P. R. Reddy, K. V. Reddy, and M. M. T. Khan,** *J. Inorg. Nucl. Chem.,* 1978, *40,* 1265.
163. **P. R. Reddy and M. H. Reddy,** *J. Chem. Soc. Dalton Trans.,* 1985, 239.
164. **P. R. Reddy, M. H. Reddy, and K. V. Reddy,** *Inorg. Chem.,* 1984, *23,* 974.
165. **M. D. Reily, K. Wilkowski, K. Shinozuka, and L. G. Marzilli,** *Inorg. Chem.,* 1985, *24,* 37.
166. **A. B. Robins,** *Chem. Biol. Interact.,* 1973, *6,* 35.
167. **F. D. Rochon, P. C. Kong, B. Coulombe, and R. Melanson,** *Can. J. Chem.,* 1980, *58,* 381.
168. **Y. Rosopulos, U. Nagel, and W. Beck,** *Chem. Ber.,* 1985, *118,* 931.
169. **G. Ruisi, M. T. Lo Giudice, and L. Pellerito,** *Inorg. Chim. Acta,* 1984, *93,* 161.
170. **K. H. Scheller, F. Hofstetter, P. R. Mitchell, B. Prijs, and H. Sigel,** *J. Am. Chem. Soc.,* 1981, *103,* 247.
171. **K. H. Scheller and H. Sigel,** *J. Am. Chem. Soc.,* 1983, *105,* 3005.
172. **K. H. Scheller, V. Scheller-Krattiger, and R. B. Martin,** *J. Am. Chem. Soc.,* 1981, *103,* 6833,
173. **W. M. Scovell and T. O'Connor,** *J. Am. Chem. Soc.,* 1977, *99,* 120.
174. **J. Y. Séguin, P. C. Kong, and M. Zador,** *Can. J. Chem.,* 1974, *52,* 2603.
175. **J. Y. Séguin and M. Zador,** *Inorg. Chim. Acta,* 1976, *20,* 203.
176. **S. Shimokawa, H. Fukui, J. Sohma, and K. Hotta,** *J. Am. Chem. Soc.,* 1973, *95,* 1777.
177. **K. Shinozuka, K. Wilskowski, B. L. Heyl, and L. G. Marzilli,** *Inorg. Chim. Acta,* 1985, *100,* 141.
178. **H. Sigel and C. F. Naumann,** *J. Am. Chem. Soc.,* 1976, *98,* 730.
179. **R. B. Simpson,** *J. Am. Chem. Soc.,* 1964, *86,* 2059.
180. **M. M. Singh, Y. Rosopulos, and W. Beck,** *Chem. Ber.,* 1983, *116,* 1364.
181. **T. Sorrell, L. A. Epps, T. J. Kistenmacher, and L. G. Marzilli,** *J. Am. Chem. Soc.,* 1977, *99,* 2173.
182. **I. Sovago and R. B. Martin,** *Inorg. Chem.,* 1980, *19,* 2868.
183. **V. K. Srivastava,** *Indian J. Biochem.,* 1969, *6,* 149.
184. **S. Suzuki, W. Mori, and A. Nakahara,** *Bioinorg. Chem.,* 1974, *3,* 281.
185. **R. S. Taylor and H. Diebler,** *Bioinorg. Chem.,* 1976, *6,* 247.
186. **S. E. Taylor, E. Buncel, and A. R. Norris,** *J. Inorg. Biochem.,* 1981, *15,* 131.
187. **V. Theodorou and N. Hadjiliadis,** *Polyhedron,* 1985, *4,* 1283.
188. **R. S. Tobias, G. Y. H. Chu, and H. J. Peresie,** *J. Am. Chem. Soc.,* 1975, *97,* 5305.
189. **A. T. Tu and C. G. Friederich,** *Biochemistry,* 1968, *7,* 4367.
190. **A. T. Tu and J. A. Reinosa,** *Biochemistry,* 1966, *5,* 3375.

191. **S. M. Wang and N. C. Li,** *J. Am. Chem. Soc.*, 1968, *90*, 5069.
192. **T. H. Wirth and N. Davidson,** *J. Am. Chem. Soc.*, 1964, *86*, 4325.
193. **T. Yamane and N. Davidson,** *J. Am. Chem. Soc.*, 1961, *83*, 2599.
194. **T. Yamane and N. Davidson,** *Biochim. Biophys. Acta*, 1962, *55*, 609.
195. **T. Yokono, S. Shimokawa, and J. Sohma,** *J. Am. Chem. Soc.*, 1975, *97*, 3827.
196. **A. M. Yurkevich, E. G. Chauser, and I. P. Rudakova,** *Bioinorg. Chem.*, 1977, *7*, 315.
197. **L. G. Marzilli and P. Chalipoyil,** *J. Am. Chem. Soc.*, 1980, *102*, 873.
198. **A. Forsti, R. Laatikainen, and K. Hemminki,** *Chem-Biol., Interact.*, 1986, *60*, 143.
199. **M. Nee and J. D. Roberts,** *Biochemistry*, 1982, *21*, 4920.
200. **R. Bau, H. K. Choi, R. C. Stevens, and S. K. Shang Hnang,** *5th Int. Symp. Platinum and Other Metal Coordination Compounds in Cancer Chemotherapy*, M. Nicolini and G. Baudoli, Eds., Cleup Padua, Italy, 1987, 341.
201. **M. A. Bruck, H. J. Korte, R. Bau, N. Hadjiliadis, and B. K. Teo,** Extended X-ray absorption fine structure (EXAFS) spectroscopic analysis of 6-mercaptopurine riboside complexes of platinum (II) and palladium (II), in *Platinum, Gold and Other Metal Chemotherapeutic Agents* (ACS Symp. Ser. 209), S. J. Lippard, Ed., American Chemical Society, Washington, DC, 1983, 245.
202. **J. J. Roberts and M. P. Pera,** Action of Platinum Antitumor Drugs, in *Molecular Aspects of Anticancer Drug Action*, S. Neidle and M. J. Waring, Eds., Macmillan, London, 1983, 183.
203. **B. de Castro, C. C. Chiang, K. Wilkowski, L. G. Marzilli, and T. J. Kistenmacher,** *Inorg. Chem.*, 1981, *20*, 1835.
204. **J. D. Orbell, C. Solorzano, L. G. Marzilli, and T. J. Kistenmacher,** *Inorg. Chem.*, 1982, *21*, 3806.
205. **G. Raudaschl-Sieber, H. Schollhorn, U. Thewalt, and B. Lippert,** *J. Am. Chem. Soc.*, 1985, 107, 3591.
206. **J. L. van der Veer, H. van den Elst, and J. Reedijk,** *Inorg. Chem.*, 1987, *26*, 1536.
207. **H. Schöllhorn, G. Raudaschl-Sieber, G. Muller, U. Thewalt, and B. Lippert,** *J. Am. Chem. Soc.*, 1985, *107*, 5932.
208. **B. Lippert,** *J. Am. Chem. Soc.*, 1981, *103*, 5691.
209. **B. Lippert, G. Raudaschl, C. J. L. Lock, and P. Pilon,** *Inorg. Chim. Acta*, 1984, *93*, 43.
210. **R. Faggiani, C. J. L. Lock, and B. Lippert,** *J. Am. Chem. Soc.*, 1980, *102*, 5419.
211. **R. Faggiani, B. Lippert, C. J. L. Lock, and R. A. Speranzini,** *Inorg. Chem.*, 1982, *21*, 3216.
212. **G. Raudaschl-Siebewr, L. G. Marzilli, B. Lippert, and K. Shinozuka,** *Inorg. Chem.*, 1985, *24*, 989.
213. **R. Beyerle and B. Lippert,** *Inorg. Chim. Acta*, 1982, *66*, 141.
214. **B. Lippert,** *Inorg. Chim. Acta*, 1981, *56*, L23.
215. **J. R. Lusty, H. S. O. Chan, E. Khor, and J. Peeling,** *Inorg. Chim. Acta*, 1985, *106*, 209.
216. **J. L. van der Veer, G. L. Ligtvaet, and J. Reedijk,** *J. Inorg. Biochem.*, 1987, *29*, 217.
217. **A. T. M. Marcellis and J. Reedijk,** *Recl. Trav. Chim. Pays-Bas*, 1983, *102*, 121.
218. **R. W. Gellert and R. Bau,** X-ray structural studies of metal-nucleoside and metal nucleotide complexes, in *Metal Ions in Biological Systems*, Vol. 8, H. Sigel, Ed., Marcel Dekker, New York, 1979.
219. **J. Reedijkk, J. H. J. den Hartog, A. M. J. Fichtinger-Schepman, and A. T. M. Marcelis,** Specific binding of *cis*-platinum compounds to DNA and DNA fragments, in *Developments in Oncology*, Vol. 17, M. P. Hacker, E. B. Douple, and I. H. Krakoff, Eds., Martinus Nijoff, Boston, 1984, 39.

Additional Reviews

220. **M. E. Howe-Grant and S. J. Lippard,** Aqueous platinum(II) chemistry; binding to biological molecules, in *Metal Ions in Biological Systems*, Vol. 11, H. Sigel, Ed., Marcel Dekker, New York, 1980, 63.

221. **J. K. Barton and S. J. Lippard,** Heavy metal interactions with nucleic acids, in *Nucleic Acid-Metal Ion Interactions,* Vol. 1, T. G. Spiro, Ed., Wiley-Interscience, New York, 1980, 31.
222. **L. G. Marzilli, T. J. Kistenmacher, and C. L. Eichhorn,** Structural principles of metal ion-nucleotide and metal ion-nucleic acid interactions, in *Nucleic Acid-Metal Ion Interactions,* Vol. 1, T. G. Spiro, Ed., Wiley-Interscience, New York, 1980, 179.
223. **R. B. Martin and Y. H. Mariam,** Interactions between metal ions and nucleic bases, nucleosides, and nucleotides in solution, in *Metal Ions in Biological Systems,* Vol. 8, H. Sigel, Ed., Marcel Dekker, New York, 1979, 57.
224. **J. Reedijk,** *Pure Appl. Chem.,* 1987, *59,* 181.
225. S. J. Lippard, Ed., *Platinum, Gold and Other Metal Chemotherapeutic Agents* (ACS Symp. Ser. 209), American Chemical Society, Washington, D.C., 1983.
226. **J. P. Macquet, J. L. Butour, and N. P. Johnson,** Physiocochemical and strucutral studies of the in vitro interactions between platinum(II) compounds and DNA, in *Platinum, Gold and Other Metal Chemotherapeutic Agents* (ACS Symp. Ser. 209), S. J. Lippard, Ed., American Chemical Society, Washington, D.C., 1983, 75.
227. **I. M. Ismail and P. J. Sadler,** ^{195}Pt- and ^{15}N-NMR studies of antitumor complexes, in *Platinum, Gold and Other Metal Chemotherapeutic Agents* (ACS Symp. Ser. 209), S. J. Lippard, Ed., American Chemical Society, Washington, D.C., 1983, 171.
228. **T. J. Kistenmacher, J. D. Orbell, and L. G. Marzilli,** Conformational properties of purine and pyrimidine complexes of *cis*-platinum: inplications for platinum(II)-DNA crosslinking modes, in *Platinum, Gold and Other Metal Chemotherapeutic Agents* (ACS Symp. Ser. 209), S. J. Lippard, Ed., American Chemical Society, Washington, D.C., 1983, 191.
229. **A. P. Hitchcock, B. Lippert, C. J. L. Lock, and W. M. C. Pratt,** Platinum complexes with DNA bases, nucleotides and DNA, in *Platinum, Gold and Other Metal Chemotherapeutic Agents* (ACS Symp. Ser. 209), S. J. Lippard, Ed., American Chemical Society, Washington, D.C., 1983, 209.
230. **R. B. Martin,** Hydrolytic equilibria and N7 versus N1 binding in purine nucleosides of cis-diamminedichloroplatinum(II): palladium(II) as a guide to platinum(II) reactions at equilibrium, in *Platinum, Gold and Other Metal Chemotherapeutic Agents* (ACS Symp. Ser. 209), S. J. Lippard, Ed., American Chemical Society, Washington, D.C., 1983, 231.
231. **L. G. Marzilli,** Metal-ion interactions with nucleic acids and nucleic acid derivatives, in *Progress in Inorganic Chemistry,* Vol. 23, S. J. Lippard, Ed., John Wiley & Sons, New York, 1977, 255.
232. **M. P. Hacker, E. B. Douple, and I. H. Krakoff, Eds.,** *Developments in Oncology,* Vol. 17, Martinus Nijhoff, Boston, 1984.
233. **J. P. Caradonna and S. J. Lippard,** Chemical and Biological studies of *cis*-diamminedichloroplatinum(II) binding to DNA, in platinum coordination complexes in cancer chemotherapy, in *Developments in Oncology,* Vol. 17, M. P. Hacker, E. B. Douple, and I. H. Krakoff, Eds., Martinus Nijhoff, Boston, 1984, 14.
234. **J. P. Macquet, J. L. Butour, N. P. Johnson, H. Razaka, B. Salles, C. Vieussens, and M. Wright,** Is DNA the real target of antitumor platinum compounds?, in *Developments in Oncology,* Vol. 17, M. P. Hacker, E. B. Douple, and I. H. Krakoff, Eds., Martinus Nijhoff, Boston, 1984, 27.
235. **J. J. Roberts and M. P. Pera, Jr.,** Action of platinum antitumor drugs, in *Molecular Aspects of Anti-Cancer Drug Action,* S. Neidle and M. J. Waring, Eds., Macmillan, London, 1983.

Section 6

COMPLEXES INVOLVING NUCLEOTIDES AND OLIGONUCLEOTIDES OF THE PURINE BASES

Kenji Inagaki and Yoshinori Kidani

INTRODUCTION

This section covers platinum and palladium adducts with purine nucleotides, oligonucleotides, and DNA. The presence and absence of the 2′-OH group of the sugar does not seem to affect the final platinum (palladium) adducts with nucleotides. In the list mentioned below, bifunctional platinum compounds having *cis*-geometry, e.g., *cis*-Pt (abbreviation of *cis*-Pt $(NH_3)_2Cl_2$ or its aquated species), Pt (en)Cl_2 and so on, possess antitumor activity. *Trans*-Pt (*trans*-Pt $(NH_3)_2Cl_2$ or its aquated species) has a bifunctional nature, but it is an antitumor-inactive compound. Monofunctional platinum compounds, [Pt (dien) Cl] Cl, and [Pt $(NH_3)_3$Cl]Cl, are also antitumor-inactive compounds. Purine nucleotides mainly coordinate to Pt (Pd) atom via the N7 and the N1, i.e., the behavior does not differ from that of purine nucleosides. However, the presence of 5′-phosphate group has a tendency to promote the reactivity of the metal compounds toward purine bases and to enhance the ratio of binding at N7 to N1.[3,108h] This appears to be due to an electrostatic attraction between a negatively charged phosphate group and a positively charged metal center. In studies on reaction rate, the rateconstant of binding of *cis*-Pt to guanosine-5′-monophosphate is faster than that to guanosine-3′-phosphate. The 5′-phosphate group exists in a convenient position to bring *cis*-Pt near the N7 of the guanine bases. A similar behavior has also been observed between *cis*-Pt and oligonucleotides.[44,58]

The reaction of dinucleotides (GpG, ApG, GpA, CpG, GpC, and so on) with bifunctional platinum compounds having a *cis*-geometry is able to give a chelate between neighboring bases which can predominate in platinum adducts. *Trans*-Pt could not form such chelate because of a stereochemical reason. The reaction of d(GpG) with *cis*-Pt yields only 1:1 adduct in which the platinum atom chelates between the two N7 sites of the guanines.[47-53,77,79] Such a chelate has been believed to be a major adduct produced in the reaction with DNA.[52,77,78,80,87-89] Detailed studies about *cis*-Pt $(NH_3)_2$(d(GpG)) show that the d(GpG) has a head-to-head arrangement with a dihedral angle of about 60°, and the sugar moieties are in an anti-anti orientation.[51] Moreover the conformation of the furanose ring of the 5′ side changes from the *S*-type conformer to the *N*-type conformer upon platination.[51]

Reaction of *cis*-Pt with the dinucleotides involving guanine and cytosine yields GN7, CN3 chelates.[47,49] The chelates obtained from the reaction with d(GpC), are mixtures of several conformational isomers,[47,49] i.e., G_{anti}-C_{anti},

G_{anti}-C_{syn}, and an equilibrium mixture of G_{syn}-C_{anti} and G_{syn}-C_{syn}. Each conformational isomer has been separated by means of HPLC with a reverse-phase column and has been characterized by high-field nuclear magnetic resonance spectroscopy. The reaction between *cis*-Pt and r(ApG) results in the AN7, GN7 chelate as a major reaction product (>95%), the structure of which is almost comparable with that of the chelate with neighboring guanines (r(GpG) and d(GpG).[46] The reaction with r(GPA) yields either GN7, AN7 and GN7, AN1 chelate.[46] It should be noted that the platinum complexes mentioned in the list contain such conformational mixtures. Moreover, attention should be given to the fact that the reaction of dinucleotides with platinum compounds lacking C_2 symmetry, e.g., Pt(1*R*, 2*S*-cyclohexanediamine)Cl_2, results in diastereoisomer.[77,79]

Binding mode of bifunctional platinum compounds (*cis*-Pt and *trans*-Pt) to oligonucleotides having –GpNpG– sequence, in which both the guanines are separated by a third base, is of interest. In the complex with *cis*-Pt, the third base is bulged out as a result of chelation of *cis*-Pt to both the guanines.[90,91,100] *Trans*-Pt also binds to both guanines to form chelate.[73,61] Such chelation has been noted as a possible binding mode of *trans*-Pt to DNA.[102]

Reaction of *cis*-Pt with self-complementary tetra- and hexanucleotides, e.g., d(CpCpGpG),[43] d(ApGpGpCpCpT),[67]d(TpGpGpCpCpA),[65] and so on, yields complexes with an intrastand cross-link between the adjacent guanines via N7 sites. In these studies, chelation of *cis*-Pt to the adjacent guanines leads to a disruption of the double helix.

In order to elucidate a distortion in the structure of oligonucleotide induced by chelation of *cis*-Pt to the adjacent guanines, duplex formation between platinated oligonucleotide strand and the complementary strand has been examined.[71,80,82] Such studies have been performed using large oligonucleotides such as octa-, deca-, and undeca-nucleotides. When *cis*-Pt was allowed to react with oligonucleotide containing –GpG– sequence, e.g., d(TpCpTpCpGpGpTpCpTpC), chelates, being cross-linked through the N7 positions of the centrally positioned guanines, are produced in a high yield.[80,82] To the platinated decanucleotide, addition of the complementary strand, d(GpApGpApCpCpGpApGpA), results in duplex formation, being shown by the appearance of imino-proton resonances arising from the central GC base pairs.[80,82] The interesting observation suggests that chelation of *cis*-Pt to an adjacent guanine site in DNA tends to destabilize the duplex structure (decrease in melting temperature), but hydrogen bonding between the double strands may be retained even in the neighborhood of platinum binding sites.

Approaches to determine the platinum binding sites in DNA have mainly been performed by a degradation of platinum-modified DNA using enzymes (or acid) and by the subsequent indentification of the digested fragments by means of electrophoresis and high-performance liquid chromatography (HPLC).

1. Electrophoresis in combination with restriction enzymatic digestion.[96-100,102]

2. Electrophoresis and/or HPLC in combination with endo- and exonuclease digestion.[52,77,78,90-93,101]
3. Electrophoresis and HPLC in combination with acid degradation.[94,95]
4. Other (nuclear magnetic resonance, X-ray (t-RNA)).[80,87-89]

The following platinum adducts have been identfied from platinum modified DNA and polynucleotides:

1. Adduct of *cis*-Pt (or its analogues such as $Pt(en)Cl_2$) with intrastrand cross-link between two adjacent guanines (–GpG–) sequence.[52,77,78,80,87-89]
2. Adduct of *cis*-Pt with intrastrand cross-link between adenine and guanine residues (–ApG– sequence, not –GpA– sequence).[52,89,92]
3. Adduct of cis-Pt with instrastrand cross-link between two guanines separated by a third base (–GpNpG– sequences, N is any nucleotide) and or adduct of *cis*-Pt with interstrand cross-link between two guanines.[52,90,91,100]
4. Adduct formed between *cis*-Pt and one guanine residue (monofunctional binding).[52,90,93]
5. Adduct of *cis*-Pt with intrastrand cross-link between the terminal bases having –ApNpG– sequences.[92]
6. Adduct of trans-Pt with intrastrand cross-link between two guanines separated by a third base (–GpNpG– sequences).[102]
7. Monofunctional adduct between trans-Pt and guanine residue.[90,93]
8. Adduct of trans-Pt with interstrand cross-link (or intrastrand cross-link) between two guanines.[90,92,93]

Some of the reviews related to this section are listed in the references. The reader should note the following reports: Reference 17 vs. 16, Reference 25 vs. 24, and Reference 13 vs. 22 and 23. The above reports contain contradictory conclusions.

TABLE INDEX

Base no.	Compound	Base
A1na	Adenosine-5′-monophosphate	AMP
A1nb	Adenosine-5′-diphosphate	ADP
A1nc	Adenosine-5′-triphosphate	ATP
G1na	Guanosine-5′-monophosphate	GMP
G1nb	Guanosine-3′-monophosphate	3GMP
G1nc	Guanosine-5′-phosphate monomethylester	GMPme
G1nd	Guanosine-5′-triphosphate	GTP
I1na	Inosine-5′-monophosphate	IMP
X1na	Xanthosine-5′-monophosphate	XMP
	Oligonucleotides	

TABLE 1

Base no.	Base	Metal	Stoichiometry	Method	Ref.
			Adenosine-5′-monophosphate		
Alna	AMP	Pd(II)	Pd(dien)(AMP)	nmr	1—4
			$[Pd(dien)]_2(\mu\text{-}AMP)$	nmr	1—4
			Pd(pmdien)(AMP)	nmr	5
			Pd(glyala)(AMP)	nmr	2
			$[Pd(glyala)]_2(\mu\text{-}AMP)$	nmr	2
			Pd(glyphe)(AMP)	nmr	2
			$[Pd(glyphe)]_2(\mu\text{-}AMP)$	nmr	2
			Pd(glytyr)(AMP)	nmr	2
			$[Pd(glytyr)]_2(\mu\text{-}AMP)$	nmr	2
			$Pd(en)(AMP)_2$	nmr	6
			Pd(en)(AMP)(AMP)	nmr	1
		Pt(II)	Pt(dien)(AMP)	nmr	12
			$[Pt(dien)]_2(\mu\text{-}AMP)$	nmr	12
			cis-$Pt(NH_3)_2Cl(AMP)$	nmr	13, 14
			cis-$Pt(NH_3)_2(OH_2)(AMP)$	nmr	14
			cis-$Pt(NH_3)_2(AMP)_2$	ir	15
			cis-$Pt(NH_3)_2(AMP)_2$	nmr	16, 14
			[*cis*-$Pt(NH_3)_2Cl]_2(\mu\text{-}AMP)$	nmr	13
			[*cis*-$Pt(NH_3)_2(AMP)]_n$, (n = 1 or 2)	nmr	17
			cis-$Pt(CH_3NH_2)_2(AMP)_2$	nmr	16
			trans-$Pt(NH_3)_2(AMP)_2$	ir	15
			$Pt(en)(OH_2)(AMP)$	nmr	16
			$Pt(en)(AMP)_2$	nmr	16
			$Pt(tn)(AMP)_2$	nmr	16
			$Pt(dmen)(AMP)_2$	nmr	16
			Pt(dmen)Cl(AMP)	nmr	16
			Pt(dmen)Br(AMP)	nmr	16
			Pt(dmen)I(AMP)	nmr	16
			Adenosine-5′-diphosphate		
Alnb	ADP	Pd(II)	Pd(glytyr)(ADP)	nmr	7
			$[Pd(glytyr)]_2(\mu\text{-}ADP)$	nmr	7
		Pt(II)	[*cis*-$Pt(NH_3)_2(ADP)]_n$, (n = 1 or 2)	nmr	17
			Adenosine-5′-triphosphate		
Alnc	ATP	Pd(II)	Pd(dien)(ATP)	nmr	1—4
			$[Pd(dien)]_2(\mu\text{-}ATP)$	nmr	1—4
			Pd(glyala)(ATP)	nmr	2
			$[Pd(glyala)]_2(\mu\text{-}ATP)$	nmr	2
			Pd(glyasp)(ATP)	nmr	8
			$[Pd(glyasp)]_2(\mu\text{-}ATP)$	nmr	8

TABLE 1 (CONTINUED)

Base no.	Base	Metal	Stoichiometry	Method	Ref.
			Pd(glyphe)(ATP)	nmr	2
			$[Pd(glyphe)]_2(\mu\text{-ATP})$	nmr	2
			Pd(glytry)(ATP)	nmr	2, 7
			$[Pd(glytyr)]_2(\mu\text{-ATP})$	nmr	2, 7
			Pd(glyhis)(ATP)	nmr	9
			$[Pd(glyhis)]_2(\mu\text{-ATP})$	nmr	9
		Pt(II)	*cis*-$Pt(NH_3)_2(OH_2)(ATP)$	nmr	14
			cis-$Pt(NH_3)_2(ATP)_2$	nmr	14
			cis-$Pt(NH_3)_2(ATP)(ATP)$	nmr	14
			[*cis*-$Pt(NH_3)_2(ATP)]_n$, (n=1 or 2)	nmr	17

Guanosine-5′-monophosphate

Base no.	Base	Metal	Stoichiometry	Method	Ref.
Glna	GMP	Pd(II)	Pd(dien)(GMP)	nmr	3, 4, 10
			$[(Pd(dien))_2(\mu\text{-GMP})]$	nmr	3, 4, 10
			Pd(pmdien)(GMP)	nmr	5
			$(Pd(pmdien))_2(\mu\text{-GMP})$	nmr	5
			Pd(glyhis)(GMP)	nmr	11
			$Pd(en)(GMP)_2$	nmr	6
		Pt(II)	$Pt(NH_3)_3(GMP)$	nmr, HPLC	18
			Pt(dien)(GMP)	nmr	12, 19
				CD, nm,	20
			cis-$Pt(NH_3)_2Cl(GMP)$	nmr	13,19, 21—24
			cis-$Pt(NH_3)_2(OH_2)(GMP)$	nmr	19, 21, 22, 24, 25
			cis-$Pt(NH_3)_2(OH)_2(GMP)$	ram, nmr	26, 27
			cis-$Pt(NH_3)_2(OH_2)(GMP)$	ram, nmr	28
			cis-$Pt(NH_3)_2(GMP)_2$	nmr	19, 21—23, 25, 29
			cis-$Pt(NH_3)_2(GMP)_2$	ram, nmr	26, 27
				ir	30, 34
				ram, nmr	28
				nmr, HPLC	18
			cis-$Pt(NH_3)_2(GMP)$	nmr	13
				nmr	24
			[*cis*-$Pt(NH_3)_2(\mu\text{-GMP})]_2$	nmr	25
			[*cis*-$Pt(NH_3)_2Cl]_2(\mu\text{-GMP})$	nmr	22, 23
			cis-$Pt(CH_3NH_2)_2(OH_2)(GMP)$	nmr	24
			cis-$Pt(CH_3NH_2)_2Cl(GMP)$	nmr	24
			cis-$Pt(CH_3NH_2)_2(GMP)$	nmr	24
			trans-$Pt(NH_3)_2Cl(GMP)$	nmr	21
			trans-$Pt(NH_3)_2(OH_2)(GMP)$	nmr	21, 25
			trans-$Pt(NH_3)_2(OH_2)(GMP)$	ram, nmr	26, 27

TABLE 1 (CONTINUED)

Base no.	Base	Metal	Stoichiometry	Method	Ref.
			trans-$Pt(NH_3)_2(GMP)_2$	nmr	21, 25, 29
			trans-$Pt(NH_3)_2(GMP)_2$	ram, nmr	26, 27
			trans-$Pt(NH_3)_2(GMP)_2$	ir	30, 31
			$PtCl_3(GMP)$	nmr	32
				ir	30, 31
			$PtCl_2(GMP)_2$	nmr	32
				ir	30, 31
			$PtCl(GMP)_3$	ir	30, 31
			$Pt(GMP)_4$	ir	30, 31
			$Pt(en)(OH_2)(GMP)$	ram, nmr	26
			$Pt(en)(GMP)_2$	ram, nmr	26
			$Pt(tn)(GMP)_2$	CD, nmr	20
			$Pt(tmdap)(GMP)_2$	nmr	19
			Pt(tmdap)Cl(GMP)	nmr	19
			$Pt(tmdap)(OH_2)(GMP)$	nmr	19
			$Pt(dad)(GMP)_2$	ir, nmr	34
			$Pt(datr)(GMP)_2$	ir, nmr	34
			$Pt(NH_3)_2(GMP)_2$	nmr	81
			$Pt(en)(GMP)_2$	nmr	81
			$Pt(dmdap)(GMP)_2$	nmr	81
			$Pt(ipa)(GMP)_2$	nmr	81
			Guanosine-3′-monophosphate		
Glnb	3GMP	Pt(II)	Pt(dien)(3GMP)	nmr	19
			cis-$Pt(NH_3)_2(3GMP)_2$	ir, nmr	34
			cis-$Pt(NH_3)_2(3GMP)_2$	nmr	19
			Guanosine-5′-monophosphate monomethylester		
Glnc	GMP-me	Pt(II)	*cis*-$Pt(NH_3)_2Cl(GMPme)$	nmr	23
			cis-$Pt(NH_3)_2(GMPme)_2$	nmr	23
			[*cis*-$Pt(NH_3)_2Cl]_2(\mu$-GMPme)	nmr	23
			$Pt(tn)(GMPme)_2$	X-ray	33
			Guanosine-5′-triphosphate		
Glnd	GTP	Pd(II)	Pd(glyhis)(GTP)	nmr	11
			Inosine-5′-monophosphate		
I1na	IMP	Pd(II)	Pd(dien)(IMP)	nmr	1, 3, 4
			$[Pd(dien)]_2(\mu$-IMP)	nmr	1, 3, 4
			Pd(pmdien)(IMP)	nmr	5
			$[Pd(pmdien)]_2(\mu$-IMP)	nmr	5
			$Pd(en)(IMP)_2$	nmr	6

TABLE 1 (CONTINUED)

Base no.	Base	Metal	Stoichiometry	Method	Ref.
			Pd(en)(IMP)(IMP)	nmr	1
		Pt(II)	*cis*-$Pt(CH_3NH_2)_2Cl(IMP)$	nmr	24
			cis-$Pt(CH_3NH_2)_2(OH_2)(IMP)$	nmr	24
			cis$(Pt(CH_3NH_2)_2(IMP)$	nmr	24
			cis-$Pt(NH_3)_2Cl(IMP)$	nmr	19
			cis-$Pt(NH_3)_2(OH_2)(IMP)$	nmr	19
			cis-$Pt(NH_3)_2(OH_2)(IMP)$	ram, nmr	37
			cis-$Pt(NH_3)_2(IMP)_2$	X-ray	35
			cis-$Pt(NH_3)_2(IMP)_2$	ir	36
			cis-$Pt(NH_3)_2(IMP)_2$	nmr	19
			cis-$Pt(NH_3)_2(IMP)_2$	ram, nmr	37
			[*cis*-$Pt(NH_3)_2(IMP)]_n$	ir	36
			trans-$Pt(NH_3)_2(IMP)_2$	ir	36
			$PtCl(IMP)_3$	ir	36
			$PtCl_2(IMP)_2$	ir	36
			$Pt(tn)(IMP)_2$	X-ray	38
			Pt(tmdap)Cl(IMP)	nmr	19
			$Pt(tmdap)(OH_2)(IMP)$	nmr	19
			cis-$Pt(NH_3)_2(GMP)(IMP)$	nmr	19
			Xanthosine-5′-monophosphate		
X1na	XMP	Pt(II)	*cis*-$Pt(NH_3)_2(XMP)Cl$	ir	39
			cis-$Pt(NH_3)_2(XMP)_2$	ir	39
			trans-$Pt(NH_3)_2(XMP)Cl$	ir	39
			trans-$Pt(NH_3)_2(XMP)_2$	ir	39
			$K[(PtCl_3(XMP)]$	ir	39
			$PtCl_2(XMP)_2$	ir	39
			$[PtCl(XMP)_3]Cl$	ir	39
			$[Pt(XMP)_4]Cl_2$	ir	39
			Oligonucleotides		
		Pt(II)	Pt(dien)[d(GpA)]	nmr	40c
			Pt(dien)[d(GpT)]	nmr	40c
			Pt(dien)[d(GpC)]	nmr	40c
			Pt(dien)[d(ApG)]	nmr	41
			Pt(dien)[d(TpG)]	nmr	41
			Pt(dien)[d(CpG)]	nmr	41
			Pt(dien)[d(CpGpT)]	nmr	42
			Pt(dien)[d(GpCpG)]	nmr	43
			Pt(dien)[d(GpApG)]	nmr	44
			Pt(dien)[d(CpCpGpG)]	nmr	43
			Pt(dien)[d(TpCpTpCpGpTpCpTpC)]	nmr	42
			$[Pt(dien)]_2$[μ-d(CpCpGpG)]	nmr	43
			$[Pt(dien)]_2$[μ-d(GpCpG)]	nmr	43
			$Pt(NH_3)_3$[r(GpG)]	HPLC, enzyme	45

TABLE 1 (CONTINUED)

Base no.	Base	Metal	Stoichiometry	Method	Ref.
			$Pt(NH_3)_3$[r(ApG)]	nmr	46
			$Pt(NH_3)_3$[r(GpA)]	nmr	46
			$Pt(NH_3)_3$[d(TpCpTpCpGpTpCpTpC)]	nmr	40b
			cis-$Pt(NH_3)_2$[r(GpG)]	nmr	47—49
			cis-$Pt(NH_3)_2$[r(GpG)]	ir, nmr	50
			cis-$Pt(NH_3)_2$[d(GpG)]	nmr	48, 49 51
			cis-$Pt(NH_3)_2$[d(pGpG)]	nmr	48, 49, 52
			cis-$Pt(NH_3)_2$[d(pGpG)]	X-ray	53
			cis-$Pt(NH)_3)_2$[rCpC)]	nmr	49
			cis-$Pt(NH_3)_2$[d(pCpC)]	nmr	49
			cis-$Pt(NH_3)_2$[r(CpG)]	nmr	49, 54
			cis-$Pt(NH_3)_2$[r(CpG)]	nmr, ir	55
			cis-$Pt(NH_3)_2$[d(pCpG)]	nmr	49, 54
			cis-$Pt(NH_3)_2[r(GpC)]_2$	uv	56
			cis-$Pt(NH_3)_2[r(GpC)]_2$	nmr, ir	55
			cis-$Pt(NH_3)_2$[r(GpC)]	nmr	47, 49
			cis-$Pt(NH_3)_2$[d(pGpC)]	nmr	49
			cis-$Pt(NH_3)_2$[d(pApG)]	nmr	52, 57
			cis-$Pt(NH_3)_2$[r(ApG)]	nmr	46
			cis-$Pt(NH_3)_2$[d(ApG)]	nmr	57
			cis-$Pt(NH_3)_2$[r(GpA)]	nmr	46
			cis-$Pt(NH_3)_2$[d(GpA)]	nmr	57
			cis-$Pt(NH_3)_2$[d(pGpA)]	nmr	57
			cis-$Pt(NH_3)_2$[d(TpApGpApT)]	nmr	57
			cis-$Pt(NH_3)_2$[d(pGpGpG)]	nmr	58
			cis-$Pt(NH_3)_2$[d(GpCpG)]	nmr	43, 59, 60
			cis-$Pt(NH_3)_2$[d(GpApG)]	nmr	44
			cis-$Pt(NH_3)_2$[d(CpGpG)]	X-ray, nmr	62, 85
			cis-$Pt(NH_3)_2$[d(CpCpGpG)]	nmr	43
			cis-$Pt(NH_3)_2$[d(pGpGpCpC)]	enzyme	63
			cis-$Pt(NH_3)_2$[d(pCpCpGpG)]	enzyme	63
			cis-$Pt(NH_3)_2$[d(CpGpCpG)]	nmr	43
			cis-$Pt(NH_3)_2$[d(GpCpGpC)]	uv, Enzyme	64
			cis-$Pt(NH_3)_2$[d(pCpGpCpG)]	nmr	43
			cis-$Pt(NH_3)_2$[(dGpCpGpC)]	nmr	43
			cis-$Pt(NH_3)_2$[d(TpGpGpCpCpA)]	nmr	65, 66
			cis-$Pt(NH_3)_2$[d(ApGpGpCpCpT)]	nmr	67
			cis-$Pt(NH_3)_2$[d(GpApTpCpCp GpGpC)]	nmr	68
			cis-$Pt(NH_3)_2$[d(GpGpTpCpGp ApCpC)]	Enzyme	69
			cis-$Pt(NH_3)_2$ [d(CpGpGpApTpCpCpG)]	Enzyme	69

TABLE 1 (CONTINUED)

Base no.	Base	Metal	Stoichiometry	Method	Ref.
			cis-$Pt(NH_3)_2$[d(TpCpTpCpGpGp TpCpTpC)]	nmr	82, 80
			cis-$Pt(NH_3)_2$[d(TpCpTpCpGpTpGp TpCpTpC)]	nmr	70
			cis-$Pt(NH_3)_2$[d(GpCpCpGpGpApTp CpGpC)]	nmr	71, 83, 84
			cis-$Pt(NH_3)_2$[r(IpI)]	nmr	72, 47, 49
			cis-$Pt(NH_3)_2$[r(ApA)]	nmr	72, 47, 49
			cis-$Pt(NH_3)_2$[d(ApTpGpG)]	nmr	86, 66
			cis-$Pt(NH_3)_2$[d(CpCpApTpGpGp)]	nmr	66
			trans-$Pt(NH_3)_2$[d(GpTpG)]	nmr	73, 74
			trans-$Pt(NH_3)_2$[d(GpCpG)]	nmr	61
			trans-$Pt(NH_3)_2$[d(ApGpGpCpCpT)]	nmr	40a
			Pt(chbma)[d(GpCpG)]	nmr	75
			Pt(en)[d(GpCpG)]	nmr	75
			Pt(en)[d(TpGpGpT)]	nmr	76
			Pt(tmen)[d(GpCpG)]	nmr	75
			Pt(*R*,*R*/*S*,*S*-dach)[d(GpG)]	uv	77
			Pt(*R*,*S*-dach)[d(GpG)]	uv	77
			Pt(*R*,*R*-dach)[d(CpCpGpG)]	Enzyme, uv	78
			Pt(*R*,*R*-dach)[d(GpGpCpC)]	Enzyme, uv	78
			Pt(1,3-dach)[d(GpG)]	uv, CD	79

REFERENCES

1. **I. Sovago and R. B. Martin,** *Inorg. Chem.,* 1980, *19,* 2868.
2. **P. I. Vestues and R. B. Martin,** *Inorg. Chim. Acta,* 1981, *55,* 99.
3. **P. I. Vestues and R. B. Martin,** *J. Am. Chem. Soc.,* 1981, *103,* 806.
4. **K. H. Scheller, S. Scheller-Krattiger, and R. B. Martin,** *J. Am. Chem. Soc.,* 1981, *103,* 6833.
5. **S. H. Kim and R. B. Martin,** *Inorg. Chim. Acta,* 1984, *93,* 11.
6. **U. K. Haring and R. B. Martin,** *Inorg. Chim. Acta,* 1983, *80,* 1.
7. **H. Kozlowski, S. Wolowiec, and B. Jezowska-Trzebiatowska,** *Biochim. Biophys. Acta,* 1979, *562,* 1.
8. **H. Kozlowski,** *Inorg. Chim. Acta,* 1977, *24,* 215.
9. **H. Kozlowski and E. Matczak-Jon,** *Inorg. Chim. Acta,* 1979, *32,* 143.
10. **F. D. Rochon, P. C. Kong, B. Coulombe, and R. Melanson,** *Can. J. Chem.,* 1980, *58,* 381.
11. **E. Matczak-Jon, B. Jezowska-Trzebiatowska, and H. Kozlowski,** *J. Inorg. Biochem.,* 1980, *12,* 143.
12. **P. C. Kong and T. Theophanides,** *Bioinorg. Chem.,* 1975, *5,* 51.
13. **G. M. Clore and A. M. Gronenborn,** *J. Am. Chem. Soc.,* 1982, *104,* 1369.
14. **M. Sarrazin, V. Peyrot, and C. Briand,** *Inorg. Chim. Acta,* 1986, *124,* 87.
15. **H. A. Tajmir-Riahi and T. Theophanides,** *Inorg. Chim. Acta,* 1983, *80,* 183.
16. **M. D. Reily and L. G. Marzilli,** *J. Am. Chem. Soc.,* 1986, *108,* 6785.
17. **R. N. Bose, R. D. Cornelius, and R. E. Viola,** *J. Am. Chem. Soc.,* 1986, *108,* 4403.
18. **A. M. J. Fichtinger-Schepman, J. L. van der Veer, P. H. M. Lohman, and J. Reedijk,** *J. Inorg. Biochem,* 1984, *21,* 103.
19. **A. T. M. Marcelis, C. Erkelens, and J. Reedijk,** *Inorg. Chim. Acta,* 1984, *91,* 129.
20. **L. G. Marzilli and P. Chalilpoyil,** *J. Am. Chem. Soc.,* 1980, *102,* 873.
21. **A. T. M. Marcelis, C. G. van Kralingen, and J. Reedijk,** *J. Inorg. Biochem.,* 1980, *13,* 213.
22. **F. J. Dijt, G. W. Canters, J. H. J. den Hartog, A. T. M. Marcelis, and J. Reedijk,** *J. Am. Chem. Soc.,* 1984, *106,* 3644.
23. **S. K. Miller and L. G. Marzilli,** *Inorg. Chem.,* 1985, *24,* 2421.
24. **M. D. Reily and L. G. Marzilli,** *J. Am. Chem. Soc.,* 1986, *108,* 8299.
25. **Y. T. Fanchiang,** *J. Chem. Soc. Dalton Trans.,* 1986, 135.
26. **G. Y. H. Chu, S. Mansy, R. E. Duncan, and R. S. Tobias,** *J. Am. Chem. Soc.,* 1978, *100,* 593.
27. **S. Mansy, G. Y. H. Chu, R. E. Duncan, and R. S. Tobias,** *J. Am. Chem. Soc.,* 1978, *100,* 607.
28. **J. R. Perno, D. Cwikel, and T. S. Spiro,** *Inorg. Chem.,* 1987, *26,* 400.
29. **M. Polissiou, V. M. T. Phan, M. St.-Jacques, and T. Theophanides,** *Inorg. Chim. Acta,* 1985, *107,* 203.
30. **H. A. Tajmir-Riahi and T. Theophanides,** *Can. J. Chem.,* 1983, *61,* 1813.
31. **H. A. Tajmir-Riahi and T. Theophanides,** *Inorg. Chim. Acta,* 1983, *80,* 223.
32. **M. Polissiou, M. T. P. Viet, M. St.-Jacques, and T. Theophanides,** *Can. J. Chem.,* 1981, *59,* 3297.
33. **L. G. Marzilli, P. Chalilpoyil, C. C. Chiang, and T. J. Kistenmacher,** *J. Am. Chem. Soc.,* 1980, *102,* 2480.
34. **K. Okamoto, V. Behnam, M. T. Phanviet, M. Polissiou, J. Y. Guathier, S. Hanessian, and T. Theophanides,** *Inorg. Chim. Acta,* 1986, *123,* L3.
35. **T. J. Kistenmacher, C. C. Chiang, P. Chalilpoyil, and L. G. Marzilli,** *J. Am. Chem. Soc.,* 1979, *101,* 1143.
36. **H. A. Tajmir-Riahi and T. Theophanides,** *Can. J. Chem.,* 1984, *62,* 1429.
37. **G. Y. H. Chu and R. S. Tobias,** *J. Am. Chem. Soc.,* 1976, *98,* 2641.
38. **T. J. Kistenmacher, C. C. Chiang, P. Chalilpoyil, and L. G. Marzilli,** *Biochem. Biophys. Res. Commun.,* 1978, *84,* 70.

39. **E. Scherer, H. A. Tajmir-Riahi, and T. Theophanides,** *Inorg. Chim. Acta,* 1984, *92,* 285.
40. **M. Nicolini and G. Bandoli, Eds.,** *5th. Int. Symp. Platinum and Other Metal Coordination Compounds in Cancer Chemotherapy,* Padua, Italy, 1987; (a) S. J. Lippard, p. 37; (b) C. J. van Garderen, C. Altona, and J. Reedijk, p. 129; (c) E. L. M. Lempers, K. Inagaki, and J. Reedijk, p. 149; (d) J. C. Chottard, p. 43.
41. **J. L. van der Veer, H. P. J. M. Noteborn, H. van den Elst, and J. Reedijk,** *Inorg. Chim. Acta,* 1987, *131,* 221.
42. **C. J. van Garderen, C. Altona, and J. Reedijk,** *Recl. Trav. Chim. Pays-Bas,* 1987, *106,* 196.
43. **A. T. M. Marcelis, J. H. J. den Hartog, G. A. van der Marel, G. Wille, and J. Reedijk,** *Eur. J. Biochem.,* 1983, *135,* 343.
44. **J. L. van der Veer, H. van den Elst, J. H. J. den Hartog, A. M. J. Fichtinger-Schepman, and J. Reedijk,** *Inorg. Chem.,* 1986, *25,* 4657.
45. **K. Inagaki, K. Kasuya, and Y. Kidani,** *Chem. Lett.,* 1983, 1345.
46. **B. V. Hemelryck, J. P. Girault, G. Chottard, P. Valadon, A. Laoui, and J. C. Chottard,** *Inorg. Chem.,* 1987, *26,* 787.
47. **J. C. Chottard, J. P. Girault, G. Chottard, J. Y. Lallemand, and D. Mansuy,** *J. Am. Chem. Soc.,* 1980, *102,* 5565.
48. **J. P. Girault, G. Chottard, J. Y. Lallemand, and J. C. Chottard,** *Biochemistry,* 1982, *21,* 1352.
49. **J. C. Chottard, J. P. Girault, E. R. Guittet, J. Y. Lallemand, and G. Chottard,** in *Platinum, Gold and Other Metal Chemotherapeutic Agents* (ACS Symp. Ser. 209), S. J. Lippard, Ed., American Chemical Society, Washington, D.C., 1983, 125.
50. **K. Okamoto, V. Behnam, and T. Theophanides,** *Inorg. Chim. Acta,* 1985, *108,* 237.
51. **J. H. J. den Hartog, C. Altona, J. C. Chottard, J. P. Girault, J. Y. Lallemand, F. A. A. M. Leeuw, A. T. M. Marcelis, and J. Reedijk,** *Nucleic Acid Res.,* 1982, *10,* 4715.
52. **A. M. J. Fichtinger-Schepman, J. L. van der Veer, J. H. J. den Hartog, P. H. M. Lohman, and J. Reedijk,** *Biochemistry,* 1985, *24,* 707.
53. **S. E. Scherman, D. Gibson, A. H. J. Wang, and S. J. Lippard,** *Science,* 1985, *230,* 412.
54. **J. P. Girault, G. Chottard, J. Y. Lallemand, F. Huguenin, and J. C. Chottard,** *J. Am. Chem. Soc.,* 1984, *106,* 7227.
55. **K. Okamoto, V. Benham, and T. Theophanides,** *Inorg. Chim. Acta,* 1987, *135,* 207.
56. **K. Inagaki and Y. Kidani,** *J. Inorg. Biochem.,* 1979, *11,* 39.
57. **F. J. Oijit and J. Reedijk,** *Recl. Trav. Chim. Pays-Bas,* 1987, *106,* 198.
58. **J. L. van der Veer, G. A. van der Marel, H. van den Elst, and J. Reedijk,** *Inorg. Chem.,* 1987, *26,* 2272.
59. **J. H. J. den Hartog, C. Altona, J. H. van Boom, A. T. M. Marcelis, G. A. van der Marel, L. J. Rinkel, G. Wille-Hazeleger, and J. Reedijk,** *Eur. J. Biochem.,* 1983, *134,* 485.
60. **A. T. M. Marcelis, J. H. J. den Hartog, and J. Reedijk,** *J. Am. Chem. Soc.,* 1982, *104,* 2664.
61. **D. Gibson and S. J. Lippard,** *Inorg. Chem.,*1987, *26,* 2275.
62. **G. Admiraal, J. L. van der Veer, R. A. G. de Graaff, J. H. J. den Hartog, and J. Reedijk,** *J. Am. Chem. Soc.,* 1987, *109,* 592.
63. **K. Inagaki, K. Kasuya, and Y. Kidani,** *Inorg. Chim. Acta,* 1984, *91,* L13.
64. **K. Inagaki and Y. Kidani,** *Inorg. Chim. Acta,* 1984, *92,* L9.
65. **J. P. Girault, J. C. Chottard, E. R. Guittet, J. Y. Lallemand, T. Huynh-Dinh, and J. Igolen,** *Biochem. Biophys. Res. Commun.,* 1982, *109,* 1157.
66. **J. C. Chottard, J. P. Girault, E. Guittet, J. Y. Lallemand, T. Huynh-Dinh, J. Igolen, J. Neumann, and S. Tran-Dinh,** *Inorg. Chim. Acta,* 1983, *79,* 249.
67. **J. P. Caradonna and S. J. Lippard,** *J. Am. Chem. Soc.,* 1982, *104,* 5793.
68. **B. V. Hemelryck, E. Guittet, G. Chottard, J. Igolen, and J. C. Chottard,** *J. Am. Chem. Soc.,* 1984, *106,* 3037.
69. **K. Inagaki and Y. Kidani,** *Inorg. Chim. Acta,* 1985, *106,* 187.
70. **J. H. J. den Hartog, C. Altona, H. van den Elst, G. A. van der Marel, and J. Reedijk,** *Inorg. Chem.,* 1985, *24,* 983.

71. **B. van Hemelrryck, E. Guittet, G. Chottard, J. P. Girault, F. Herman, T. Huynh-Dinh, J. Y. Lallemand, J. Igolen, and J. C. Chottard,** *Biochem. Biophys. Res. Commun.*, 1986, *138*, 758.
72. **J. C. Chottard, J. P Girault, G. Chottard, J. Y. Lallemand, and D. Mansuy,** *Nouv. J. Chim.*, 1978, *2*, 551.
73. **J. L. van der Veer, G. J. Ligtvoet, H. van den Elst, and J. Reedijk,** *J. Am. Chem. Soc.*, 1986, *108*, 3860.
74. **W. I. Sundquist, K. Ahmed, L. S. Hollis, and S. J. Lippard,** *Inorg. Chem.*, 1987, *26*, 1524.
75. **J. H. J. den Hartog, C. Altona, J. H. van Boom, and J. Reedijk,** *Recl. Trav. Chim. Pays-Bas*, 1984, *103*, 322.
76. **R. A. Byrd, M. F. Summers, G. Zon, C. S. Fouts, and L. G. Marzilli,** *J. Am. Chem. Soc.*, 1986, *108*, 504.
77. **K. Inagaki and Y. Kidani,** *Inorg. Chem.*, 1986, *25*, 1.
78. **K. Inagaki, K. Kasuya, and Y. Kidani,** *Chem. Lett.*, 1984, 171.
79. **K. Inagaki and Y. Kidani,** *Chem. Pharm. Bull.*, 1985, *33*, 5593.
80. **J. H. J. den Hartog, C. Altona, J. H. van Boom, and J. Reedijk,** *FEBS Lett.*, 1984, *176*, 393.
81. **J. L. van der Veer, A. R. Peters, and J. Reedijk,** *J. Inorg. Biochem.*, 1986, *26*, 137.
82. **J. H. J. den Hartog, C. Altona, J. H. van Boom, G. A. van der Marel, C. A. G. Haasnoot, and J. Reedijk,** *J. Am. Chem. Soc.*, 1984, *106*, 1528. *J. Biomol. Struct. Dynam.*, 1985, *2*, 1137.
83. **F. Herman, E. Guittet, J. Kozelka, J. P. Girault, T. Huynh-Dinh, J. Igolen, J. Y. Lallemand, and J. C. Chottard,** *Recl. Trav. Chim. Pays-Bas*, 1987, *106*, 195.
84. **J. C. Chottard,** *Recl. Trav. Chim. Pays-Bas*, 1987, *106*, 192.
85. **J. H. J. den Hartog, C. Altona, G. A. van der Marel, and J. Reedijk,** *Eur. J. Biochem.*, 1985, *147*, 371.
86. **J. M. Neumann, S. Tran-Dihn, J. P. Girault, J. C. Chottard, T. Huynh-Dinh, and J. Igolen,** *Eur. J. Biochem.*, 1984, *141*, 465.
87. **L. G. Marzilli, M. D. Reily, B.L. Heyl, C. T. McMurray, and W. D. Wilson,** *FEBS Lett.*, 1984, *176*, 389.
88. **M. D. Reily and L. G. Marzilli,** *J. Am. Chem. Soc.*, 1985, *107*, 4916.
89. **J. C. Dewan,** *J. Am. Chem. Soc.*, 1984, *106*, 7239.
90. **A. M. J. Fichtinger-Schepman, P. H. M. Lohman, and J. Reedijk,** *Nucl. Acid. Res.*,1982, 10, 5345.
91. **A. Eastman,** *Biochemistry*, 1983, *22*, 3927.
92. **A. Eastman,** *Biochemistry*, 1985, *24*, 5027.
93. **A. Eastman and M. A. Barry,** *Biochemistry*, 1987, *26*, 3303.
94. **N. P. Johnson, A. M. Mazard, J. Escalier, and J. P. Macquet,** *J. Am. Chem. Soc.*, 1985, *107*, 6376.
95. **R. O. Rahn,** *J. Inorg. Biochem.*, 1984, *21*, 311.
96. **A. D. Kelman and M. Buchbinder,** *Biochimie*, 1978, *60*, 893.
97. **H. M. Ushay, T. D. Tullius, and S. J. Lippard,** *Biochemistry*, 1981, *20*, 3744.
98. **T. D. Tullius and S. J. Lippard,** *J. Am. Chem. Soc.*, 1981, *103*, 4620; *Proc. Natl. Acad. Sci. U.S.A.*, 1982, *79*, 3489.
99. **C. L. Cohen, J. A. Ledner, W. R. Bauer, H. M. Ushay, G. Caravana, and S. J. Lippard,** *J. Am. Chem. Soc.*, 1979, *102*, 2487.
100. **A. Rahmouni and M. Leng,** *Biochemistry*, 1987, *26*, 7229.
101. **K. Inagaki, T. Ninomiya, and Y. Kidani,** *Chem. Lett.*, 1986, 233.
102. **A. L. Pinto and S. J. Lippard,** *Proc. Natl. Acad. Sci. U.S.A.*, 1985, *82*, 4616.

REVIEWS

103. **M. E. Howe-Grant and S. J. Lippard,** Aqueous platinum(II) chemistry; binding to biological molecules, in *Metal Ions in Biological Systems*, Vol. 11, H. Sigel, Ed., Marcel Dekker, New York, 1980, 63.

104. **J. K. Barton and S. J. Lippard,** Heavy metal interactions with nucleic acids, in *Nucleic Acid-Metal Ion Interactions,* Vol. 1, T. G. Sprio, Ed., Wiley-Interscience, New York, 1980, 31.
105. **L. G. Marzilli, T. J. Kistenmacher, and C. L. Eichhorn,** Structural principles of metal ion-nucleotide and metal ion-nucleic acid interactions, in *Nucleic Acid-Metal Ion Interactions,* Vol. 1, T. G. Spiro, Ed., Wiley-Interscience, New York, 1980, 179.
106. **R. W. Gellert and R. Bau,** X-ray structural studies of metal nucleoside and metal nucleotide complexes, in *Metal Ions in Biological Systems,* Vol. 8, H. Sigel, Ed., Marcel Dekker, New York, 1979, 1.
107. **R. B. Martin and Y. H. Mariam,** Interactions between metal ions and nucleic bases, nucleosides, and nucleotides in solution, in *Metal Ions in Biological Systems,* Vol. 8, H. Sigel, Ed., Marcel Dekker, New York, 1979, 57.
108. **S. J. Lippard, Ed.,** *Platinum, Gold and Other Metal Chemotherapeutic Agents* (ACS Symp. Ser. 209), American Chemical Society, Washington, D.C., 1983.
108a. **J. J. Robert and M. F. Pera, Jr.,** DNA as a Target for Anticancer Coordination Compounds, 3.
108b. **T. D. Tullius, H. M. Ushay, C. M. Merkel, J. P. Caradonna, and S. J. Lippard,** Structural Chemistry of Platinum-DNA Adducts, 51.
108c. **J. P. Macquet, J. L. Butour, and N. P. Johnson,** Physicochemical and Structural Studies of the In Vitro Interactions between Platinum(II) Compounds and DNA, 75.
108d. **J. C. Chottard, J. P. Girault, E. R. Guittet, J. Y. Lallemand, and G. Chottard,** Platinum-Oligonucleotide Structures and Their Relevance to Platinum-DNA Interaction, 125.
108e. **I. M. Ismail and P. J. Sadler,** ^{195}Pt- and ^{15}N-NMR Studies of Antitumor Complexes, 171.
108f. **T. J. Kistenmacher, J. D. Orbell, and L. G. Marzilli,** Conformational Properties of Purine and Pyrimidine Complexes of cis-Platinum: Implications for Platinum(II)-DNA Crosslinking Modes, 191.
108g. **A. P. Hitchcock, B. Lippert, C. J. L. Lock, and W. M. C. Pratt,** Platinum Complexes with DNA Bases, Nucleotides, and DNA, 209.
108h. **R. B. Martin,** Hydrolytic Equilibria and N7 Versus N1 Binding in Purine Nucleosides of cis-Diamminedichloro-platinum(II): Palladium(II) as a Guide to Platinum(II) Reactions at Equilibrium, 231.
108i. **M. A. Bruck, H. J. Korte, R. Bau, N. Hadjiliadis, and B. K. Teo,** Extended X-ray Absorption Fine Structure (EXAFS) Spectroscopic Analysis of 6-Mercaptopurine Riboside Complexes of Platinum(II) and Palladium(II), 245.
109. **L. G. Marzilli,** Metal complexes of nucleic acid derivatives and nucleotides: binding sites and structures, in *Metal Ion in Genetic Information Transfer,* Elsevier, New York, 1981, 47.
110. **L. G. Marzilli,** Metal-ion interaction with nucleic acids and nucleic acid derviatives, in *Progress in Inorganic Chemistry,* Vol. 23, S. J. Lippard, Ed., John Wiley & Sons, 1977, 255.
111. **R. B. Martin,** *Acc. Chem. Res.,* 1985, *18,* 32.
112. The Proceedings of the Third International Symposium on Platinum Coordination Complexes in Cancer Chemotherapy (Dallas, Texas, U.S.A., 1976), *J. Clin. Hematol. Oncol.,* 7, 1977.
113. **M. P. Hacker, E. B. Douple, and I. H. Krakoff, Eds.,** *Developments in Oncology,* Vol. 17, Martinus Nijhoff, Boston, 1984.
113a. **J. P. Caradonna and S. J. Lippard,** Chemical and Biological Studies of cis-Diamminedichloroplatinum(II) Binding to DNA, 14.
113b. **J. P. Macquet, J. L. Butour, N. P. Johnson, H. Razaka, B. Salles, C. Vieussens, and M. Wright,** Is DNA the Real Target of Antitumor Platinum Compounds?, 27.
113c. **J. Reedijk, J. H. J. den Hartog, A. M. J. Fichtinger-Schepman, and A. T. M. Marcelis,** Specific Binding of cis-Platinum Compounds to DNA and DNA Fragments, 39.

APPENDIX

On the following pages are included structures of some of the major bases and their derivatives. In many cases other tautomeric forms have been shown to exist, and the actual structure of the base is dependent on a number of factors including the solvent, pH, etc.

The structures are listed in alphabetical order based on their abbreviations indicated in the main text.

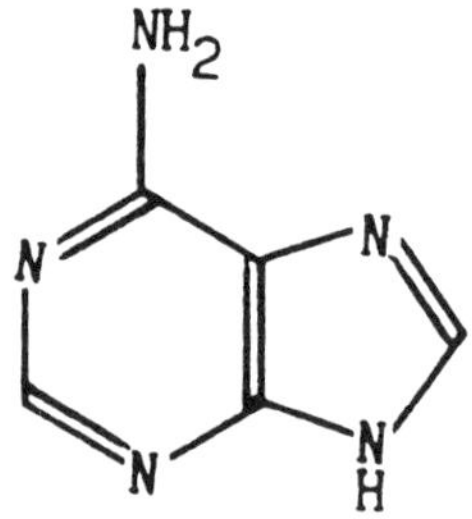

A1

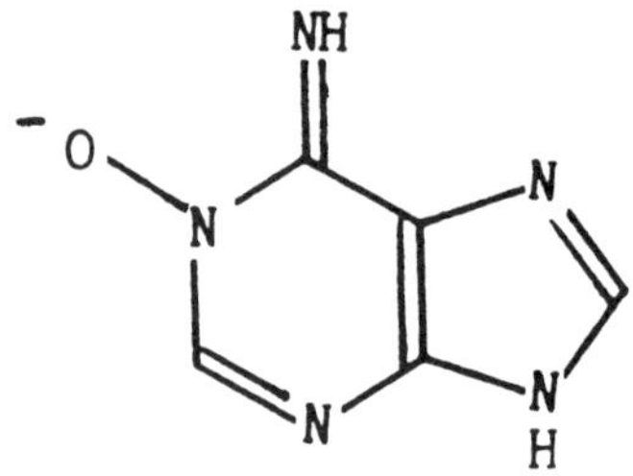

A2

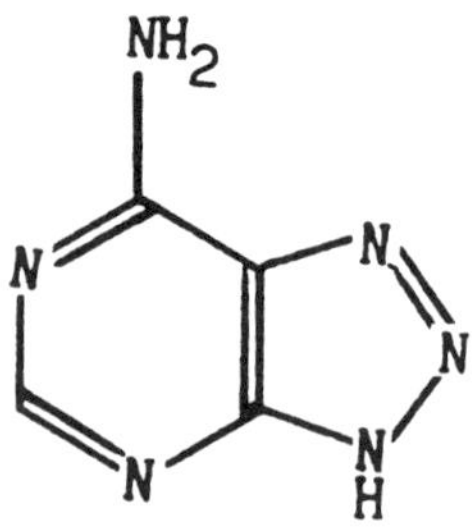

A3

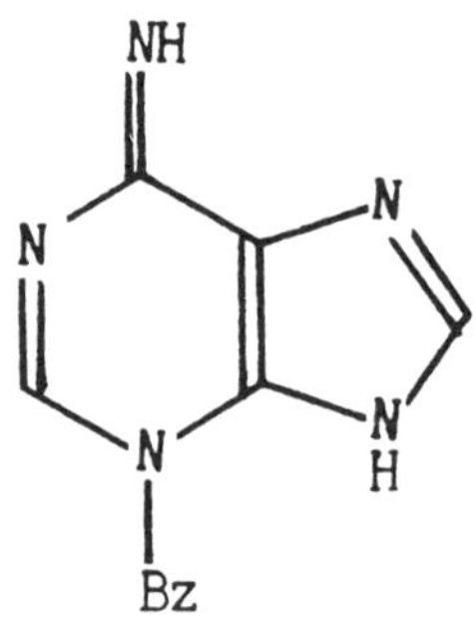

A4

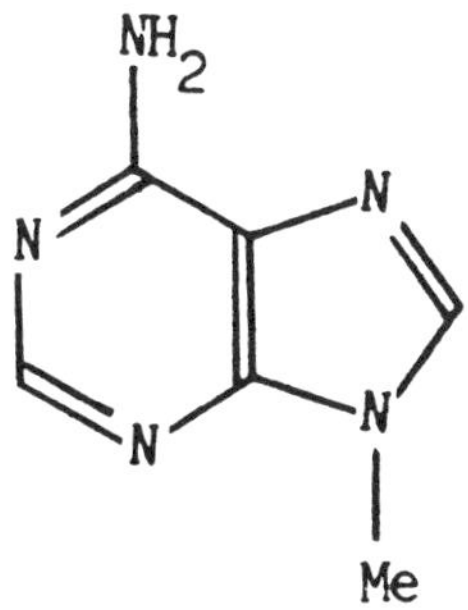

A7

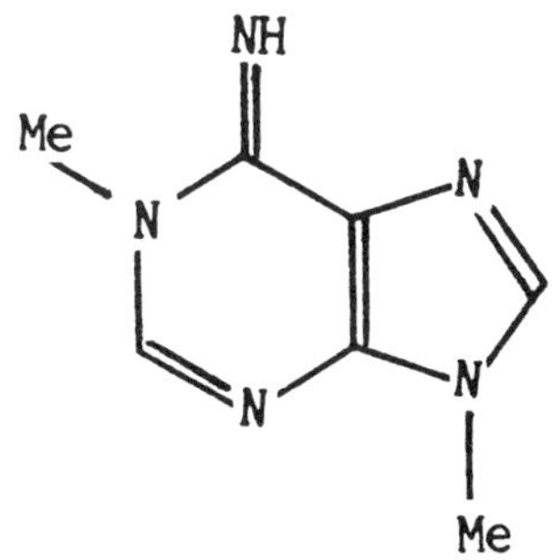

A8

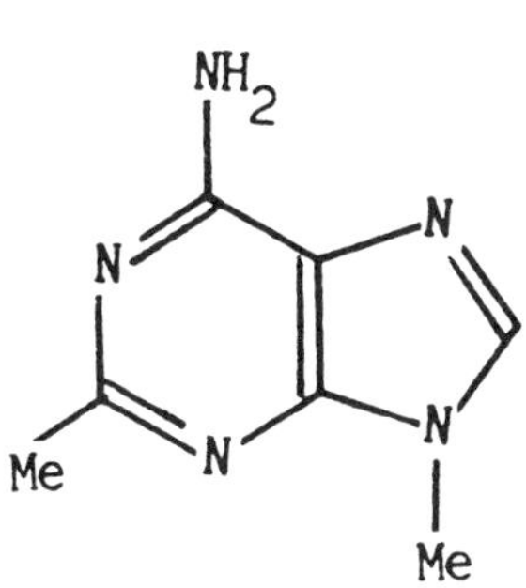

A9

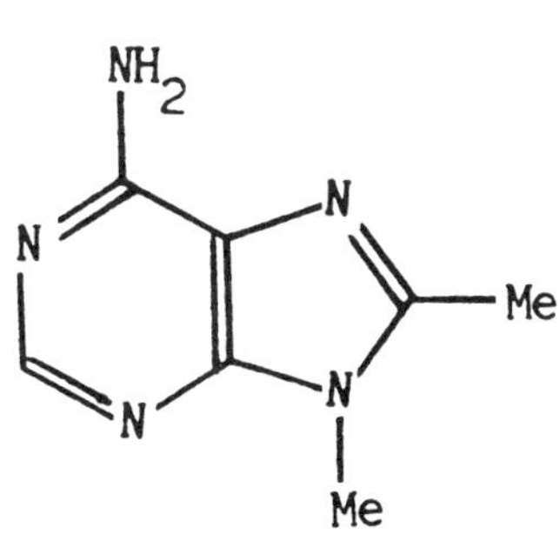

A10

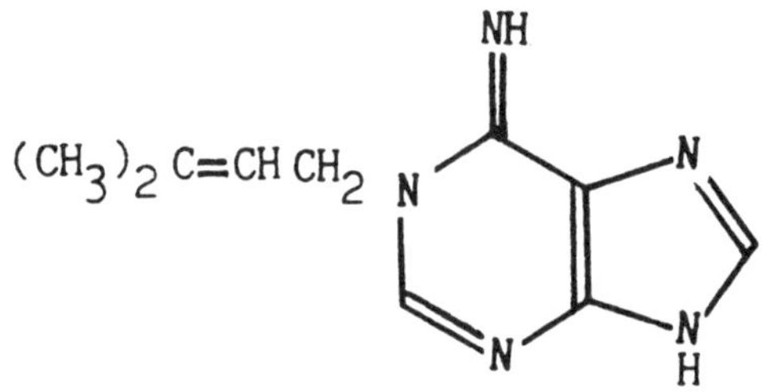
NH
$(CH_3)_2$ C=CH CH_2
N
N
N
N
H

A11

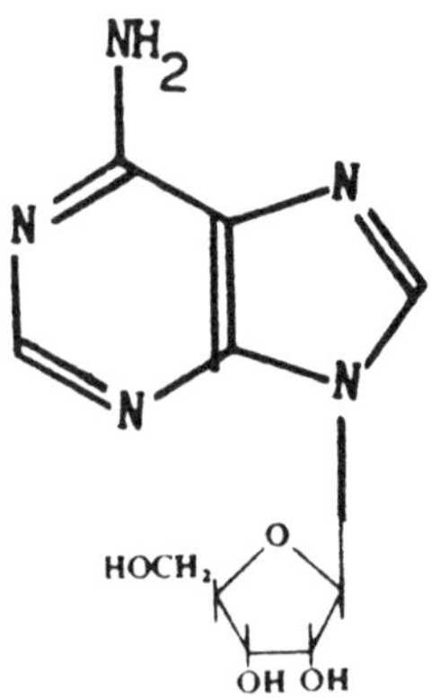
NH_2
N
N
N
N
HOCH₂
O
OH OH

A1n

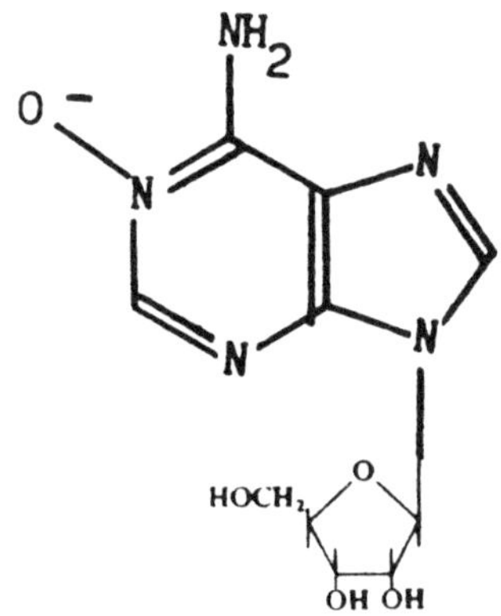
NH_2
O⁻
N
N
N
N
HOCH₂
O
OH OH

A2n

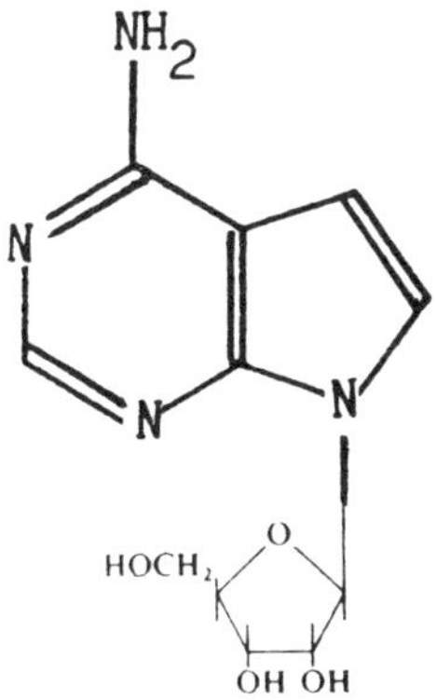

A3n

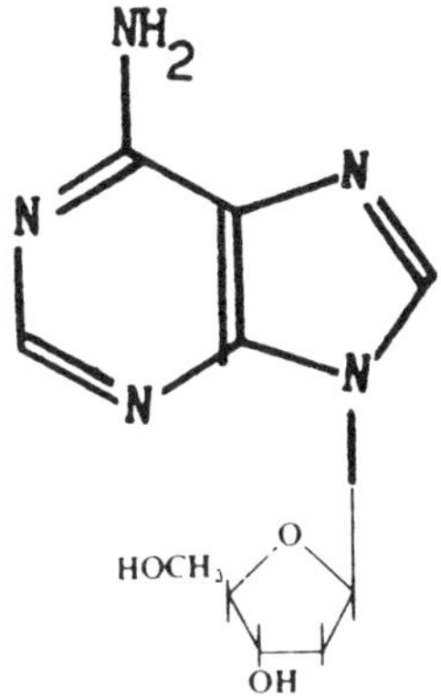

A4n

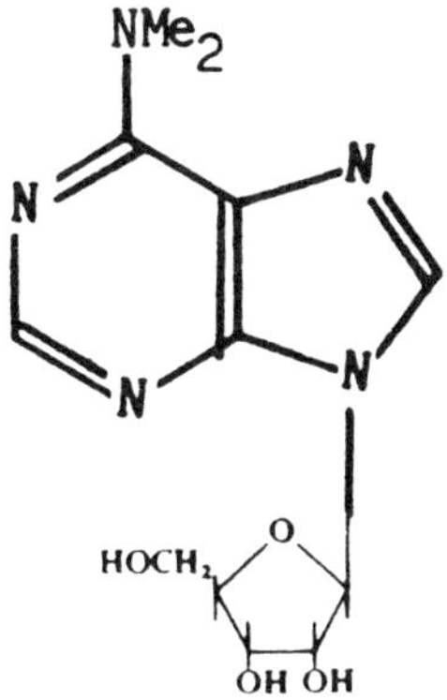

A5n

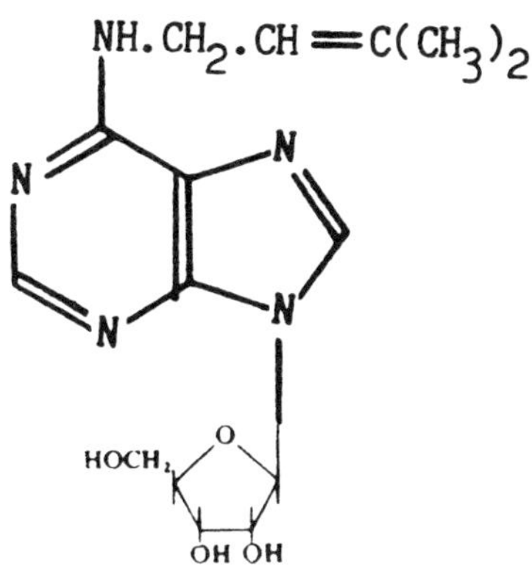

A7n

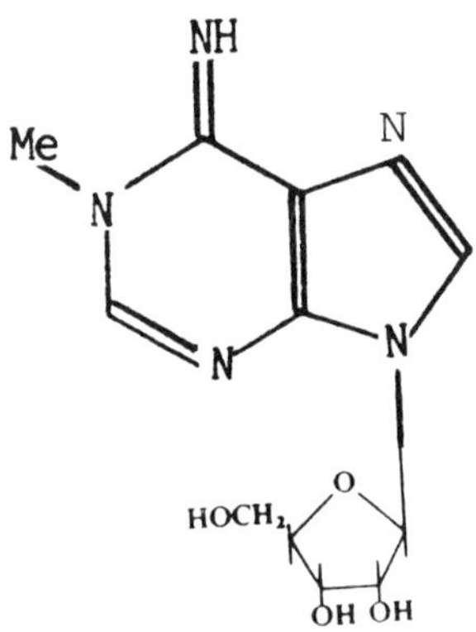

A8n

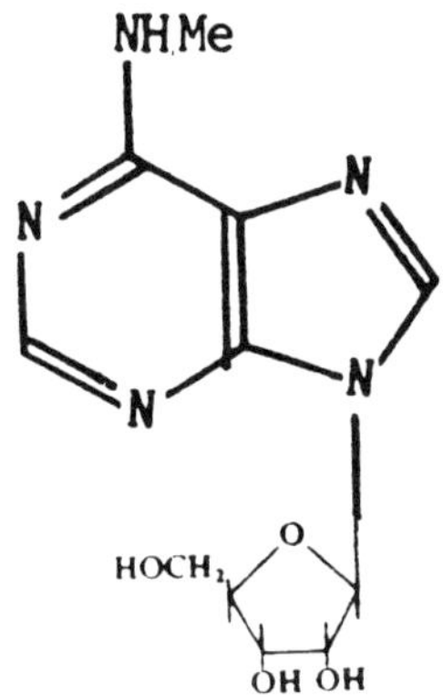

A9n

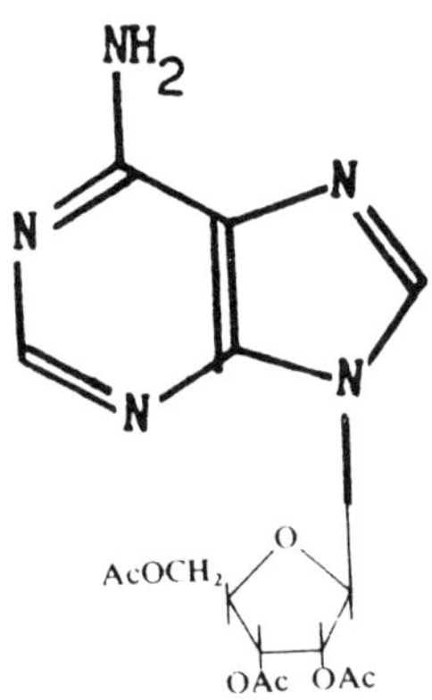

A12n

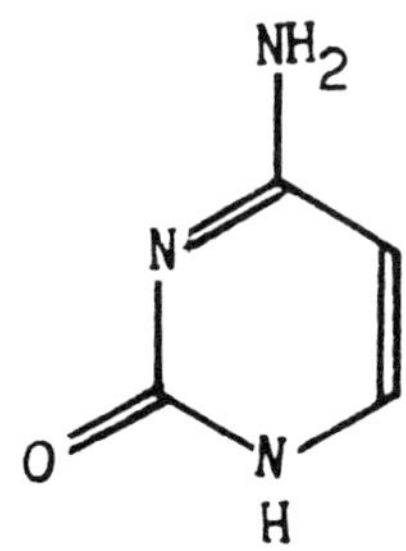

C1

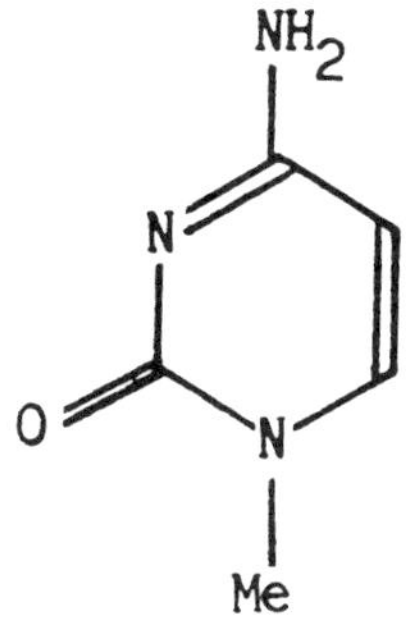

C2

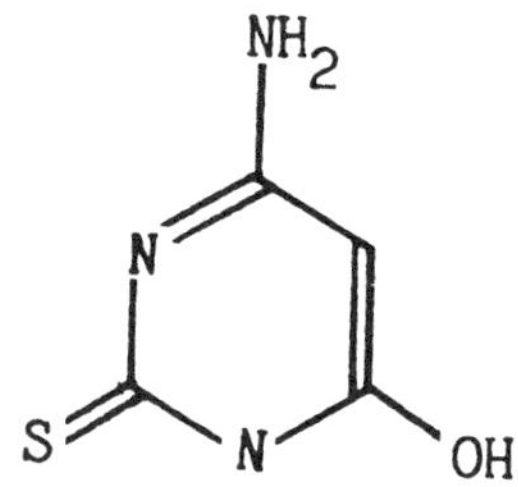

C3

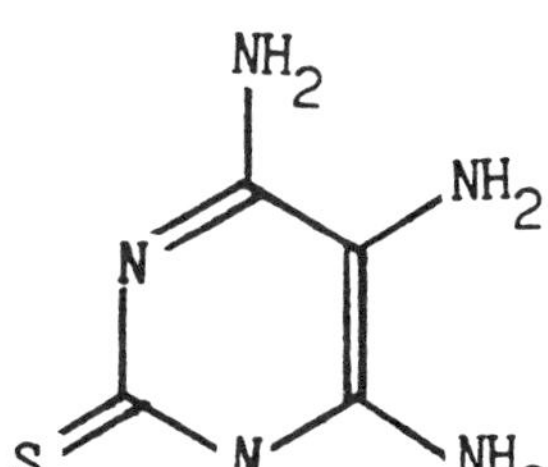

C6

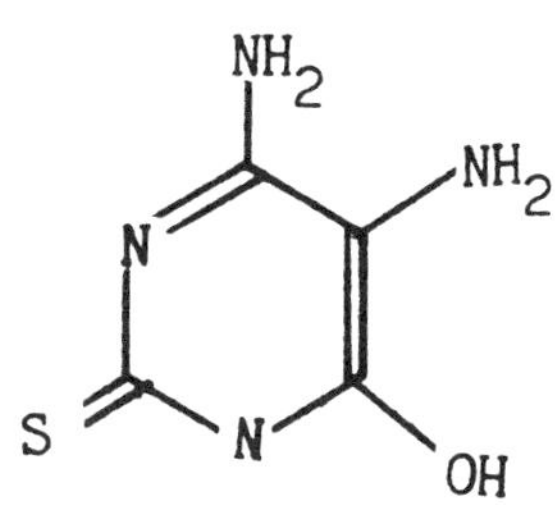

C7

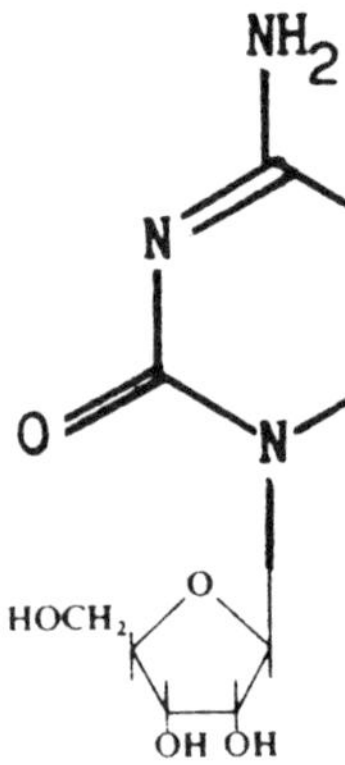
NH2
N
O
N
HOCH2
O
OH OH

C1n

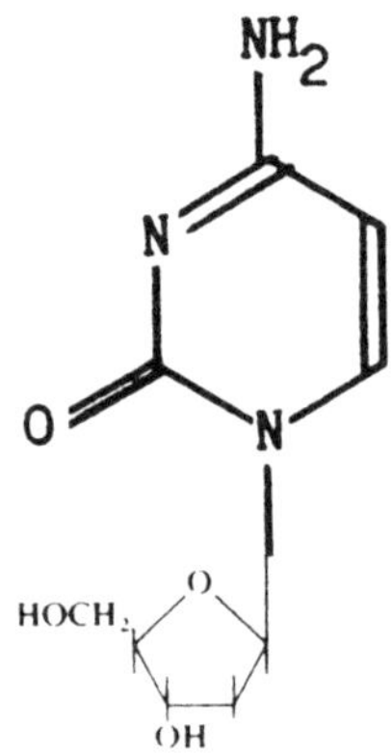
NH2
N
O
N
HOCH2
O
OH

C3n

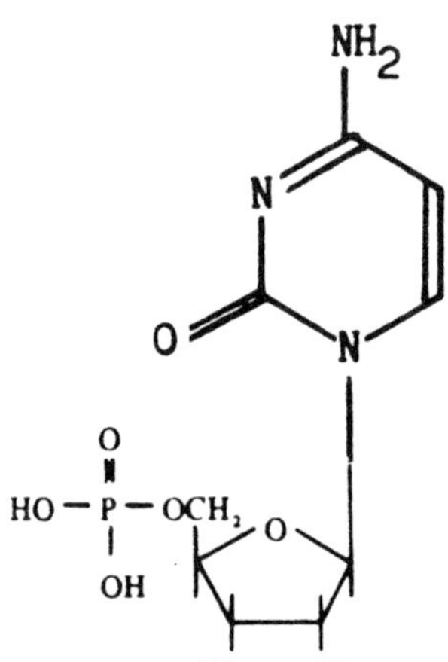
NH2
N
O
N
O
HO — P — OCH2
OH
O
HO OH

C1na

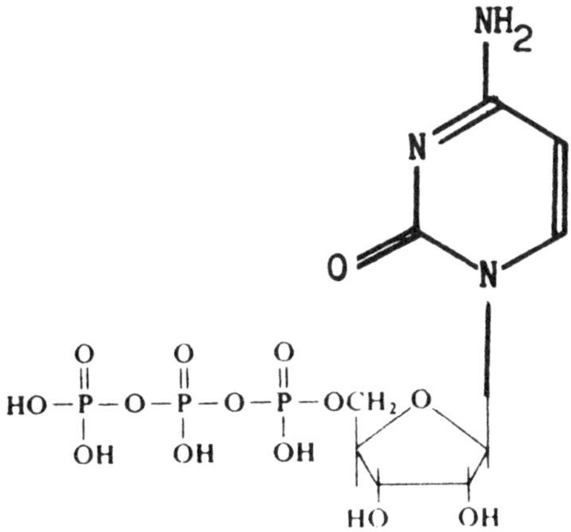

Clnc

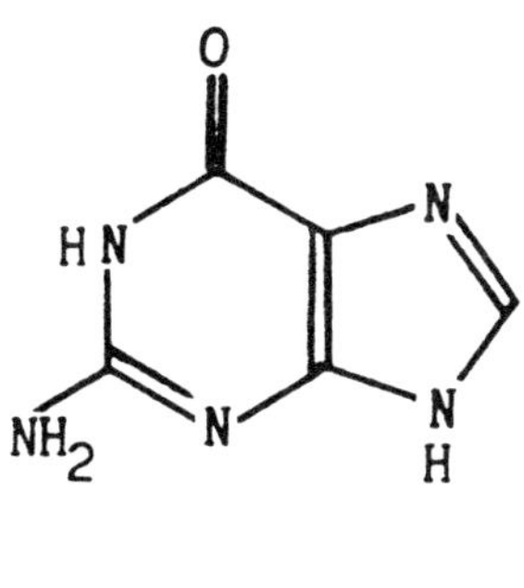

Gl

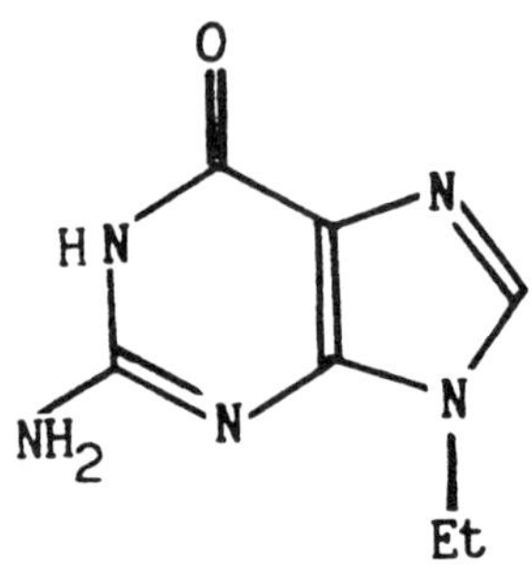

G2

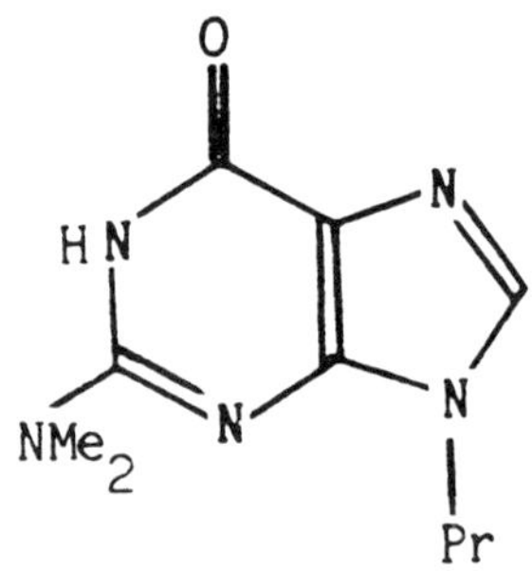

G6

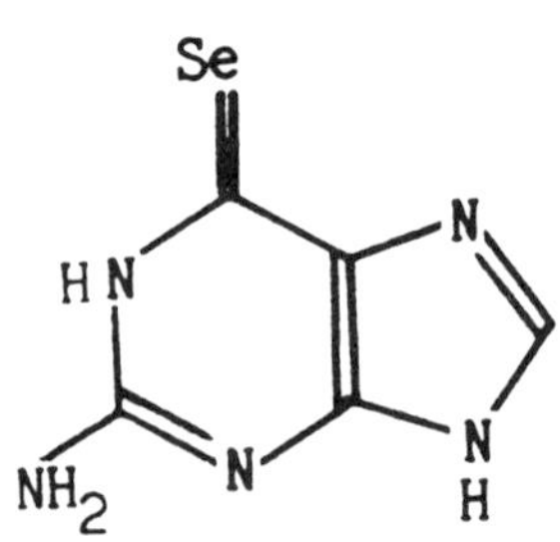

G9

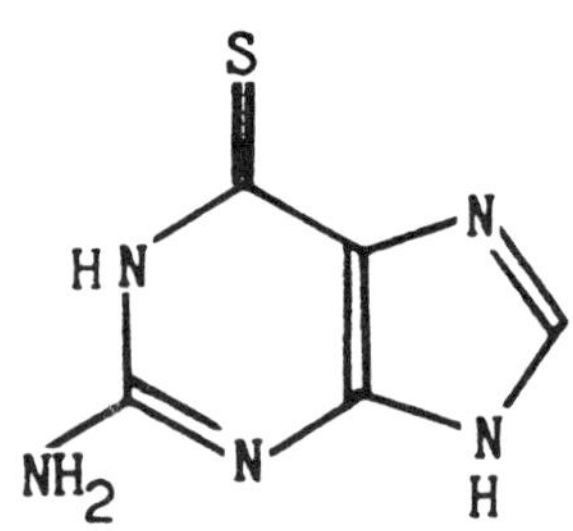

G10

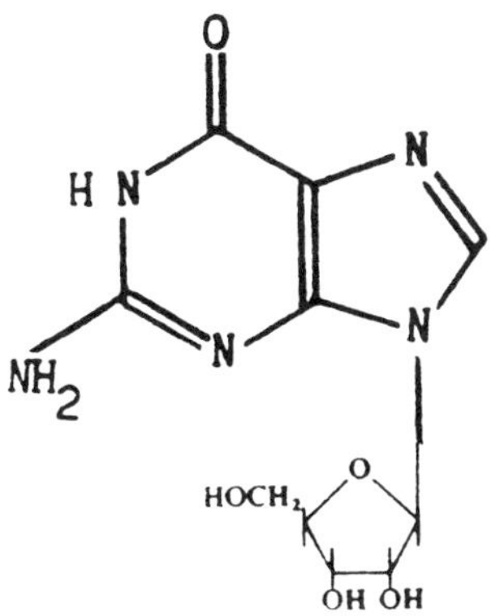
O
H N
N
N
N
NH2
HOCH2
O
OH OH

G1n

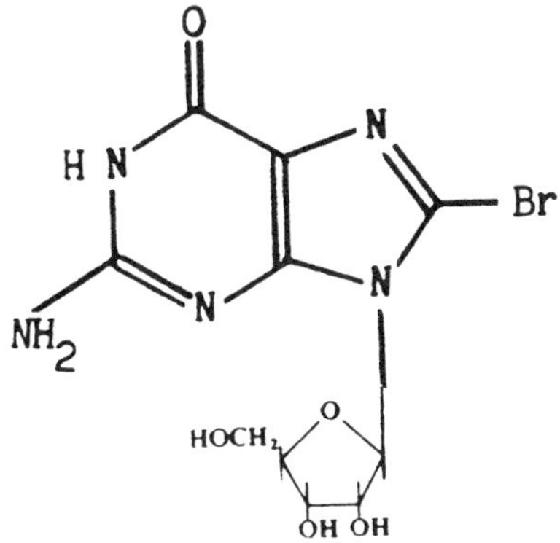
O
H N
N
Br
N
N
NH2
HOCH2
O
OH OH

G2n

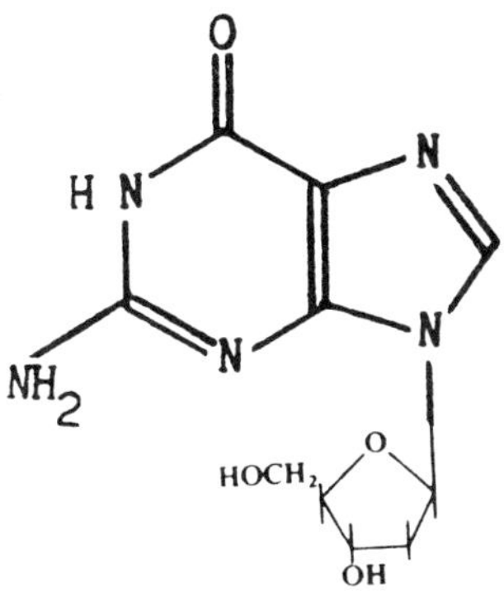
O
H N
N
N
N
NH2
HOCH2
O
OH

G3n

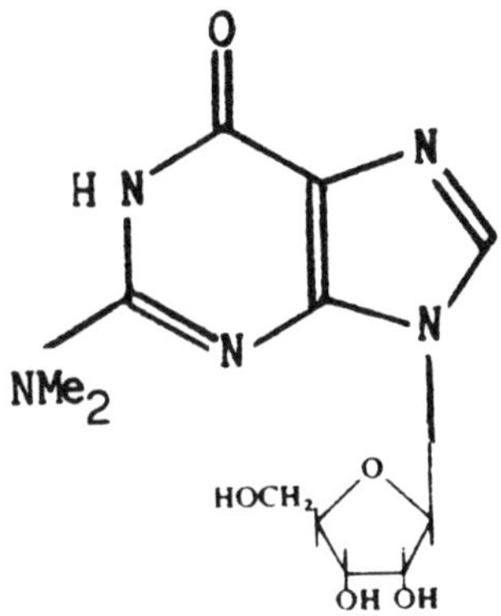

G4n

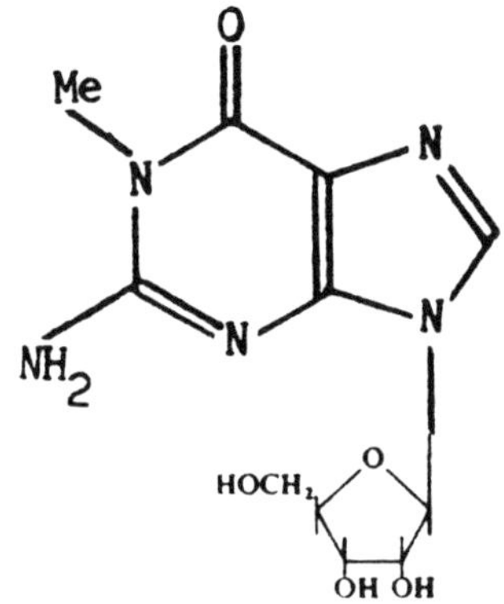

G5n

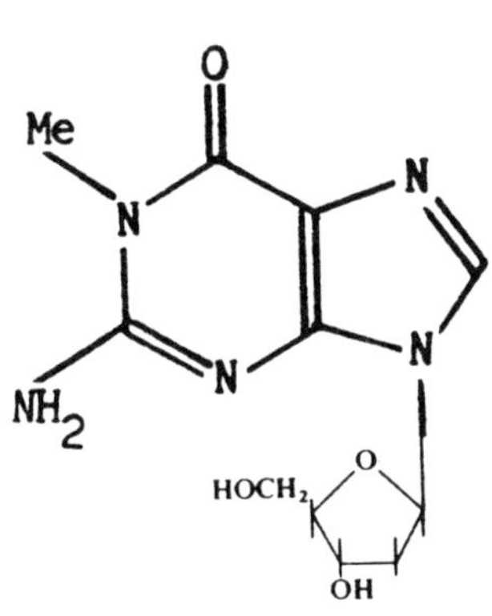

G6n

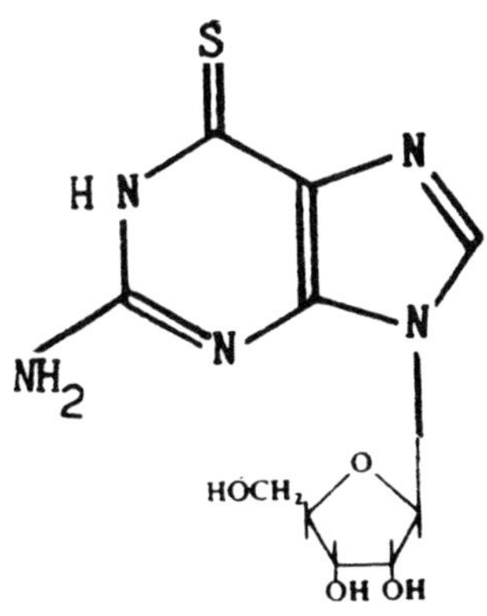

G9n

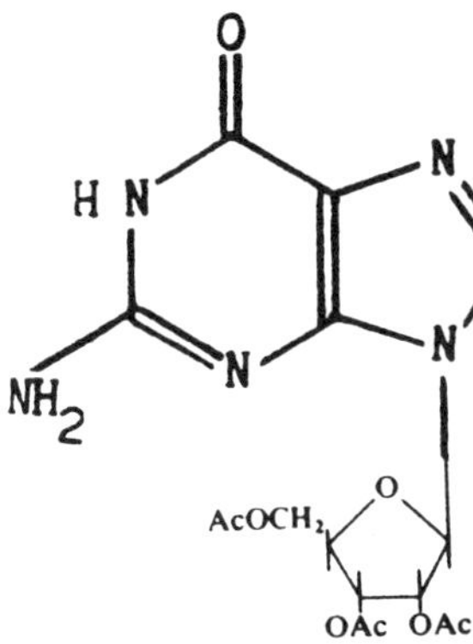

G1ln

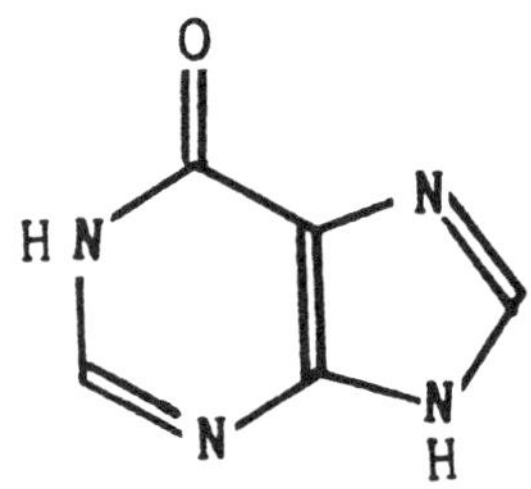

H1

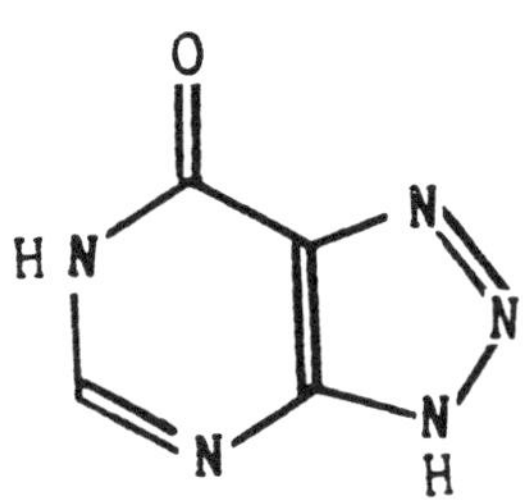

H2

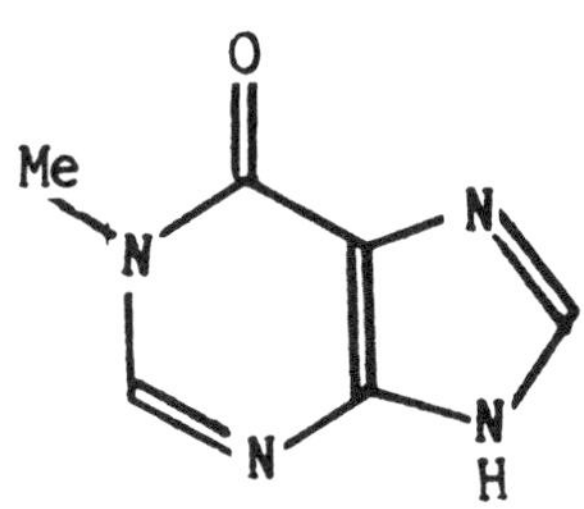

H3

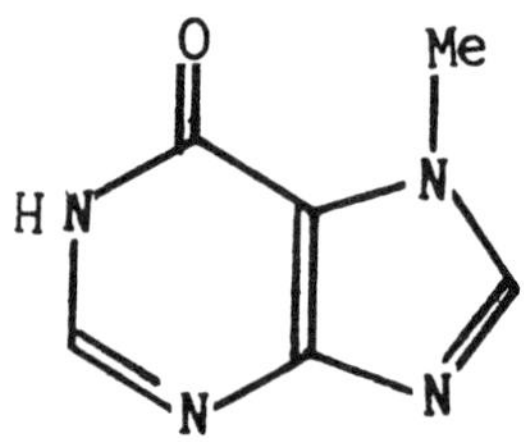

H4

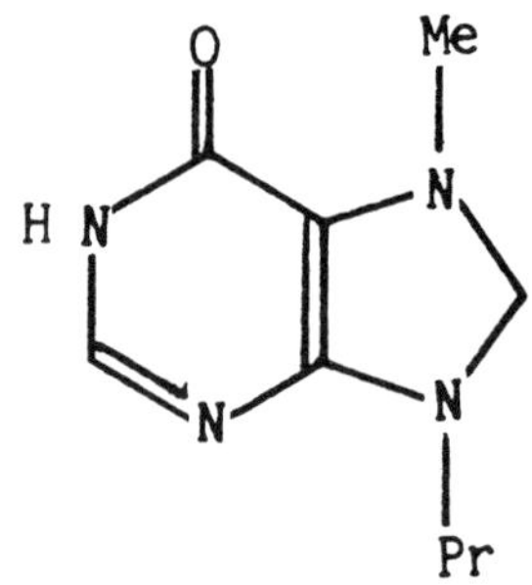

H5

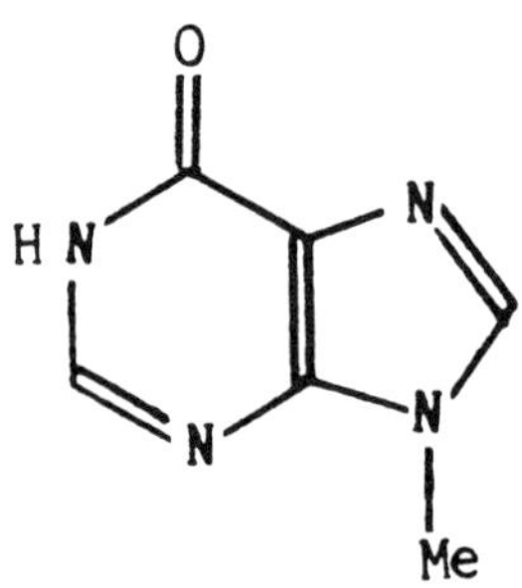

H6

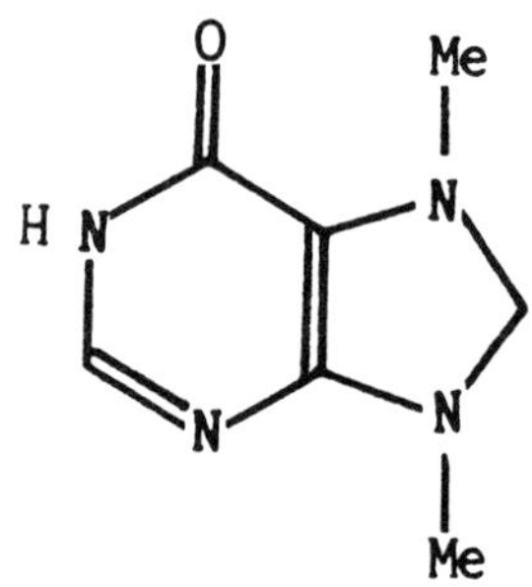

H7

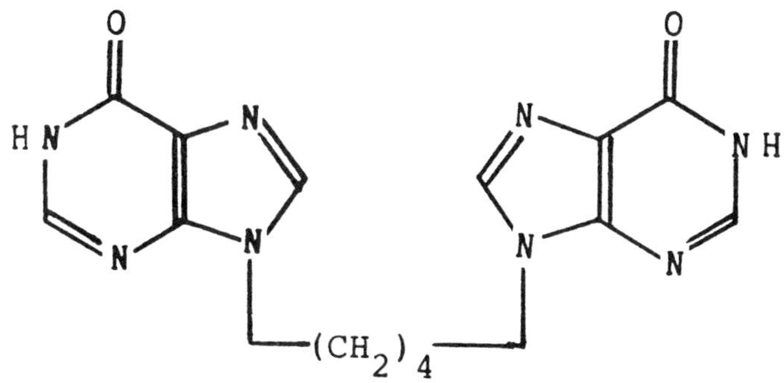

H8

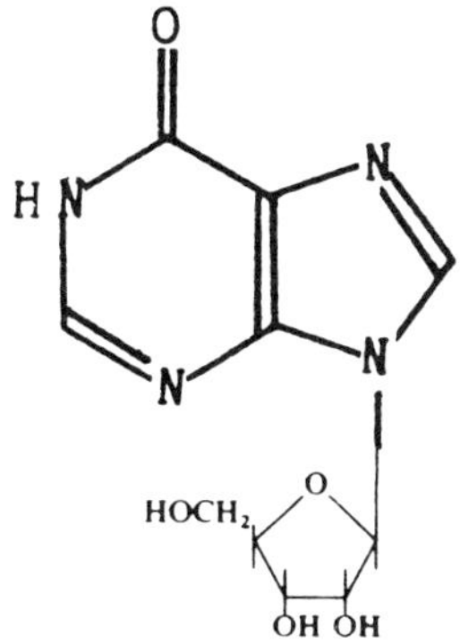

I1n

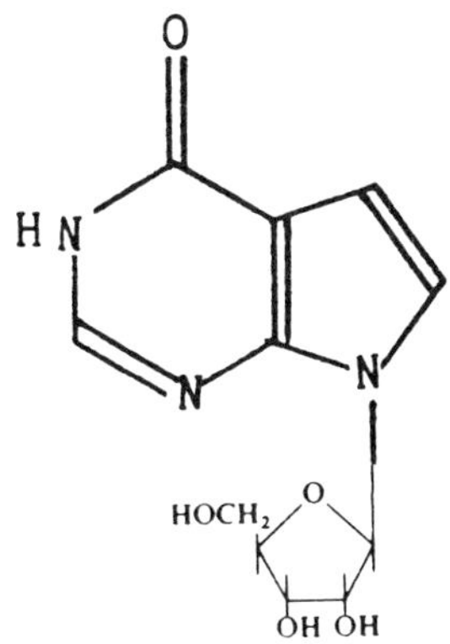

I2n

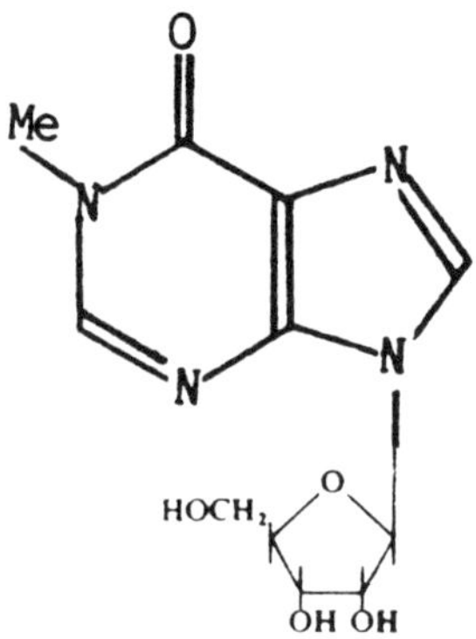

I4n

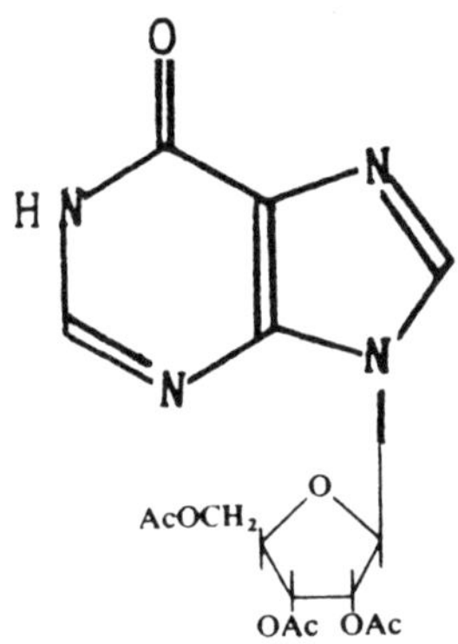

I7n

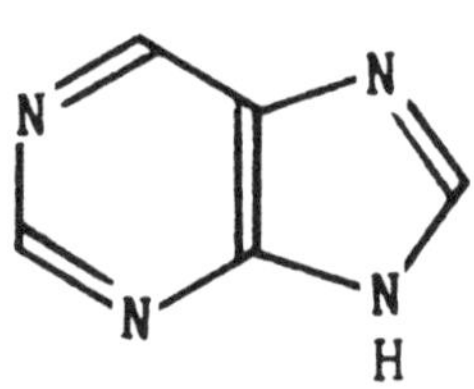

P1

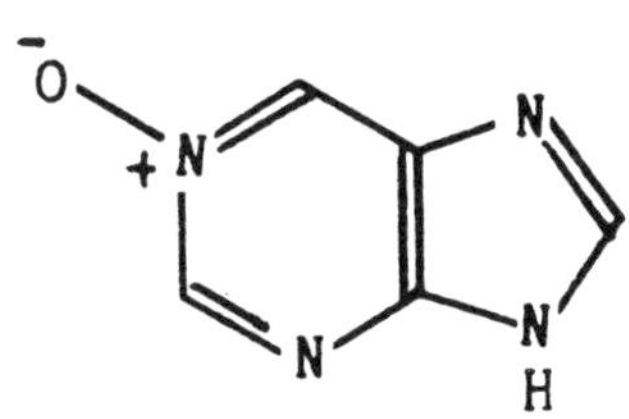

P2

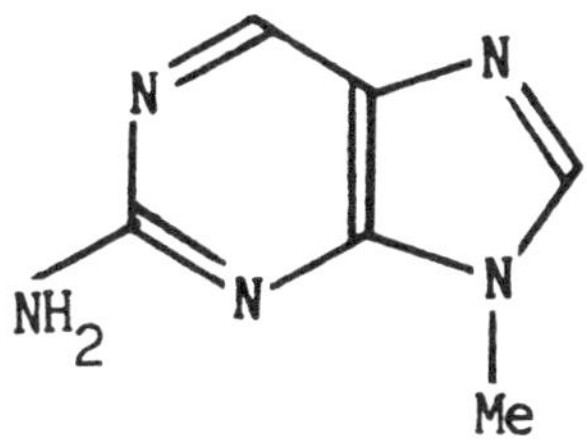

P3

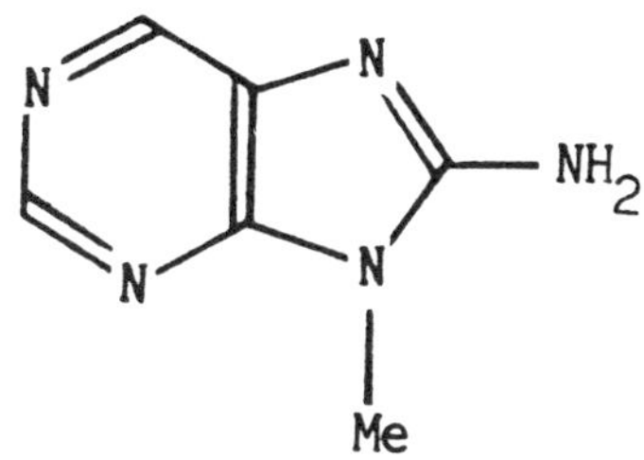

P4

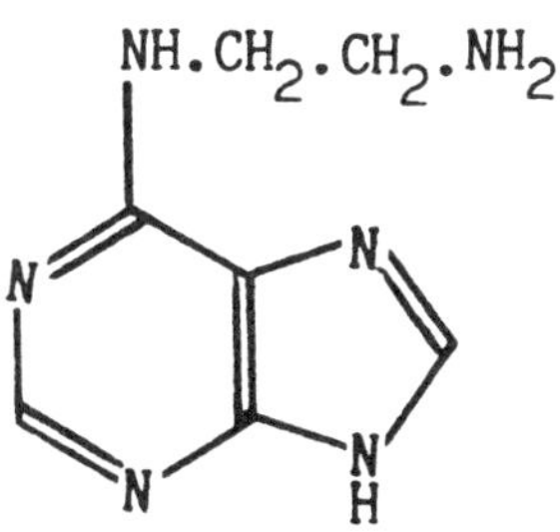

P6

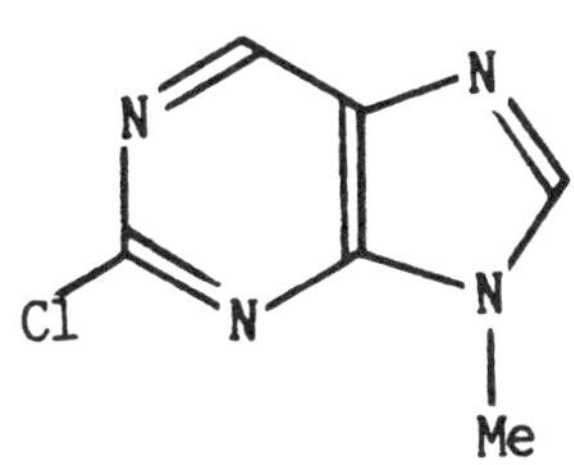

P8

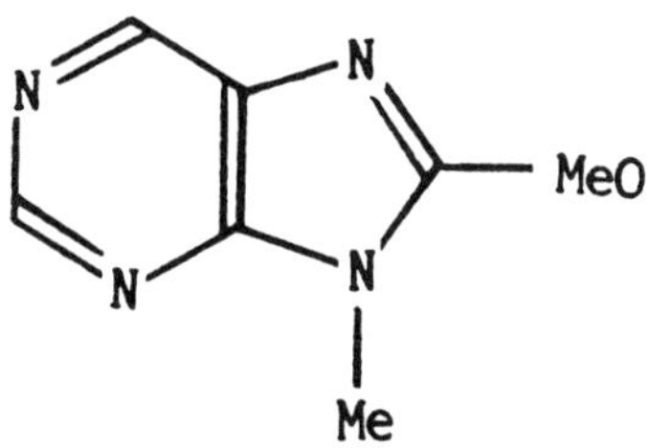

P13

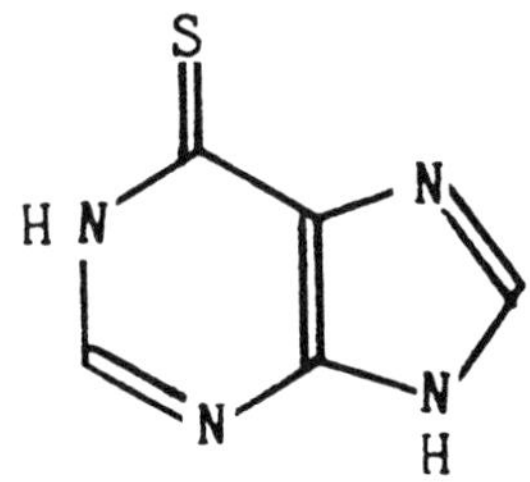

P18

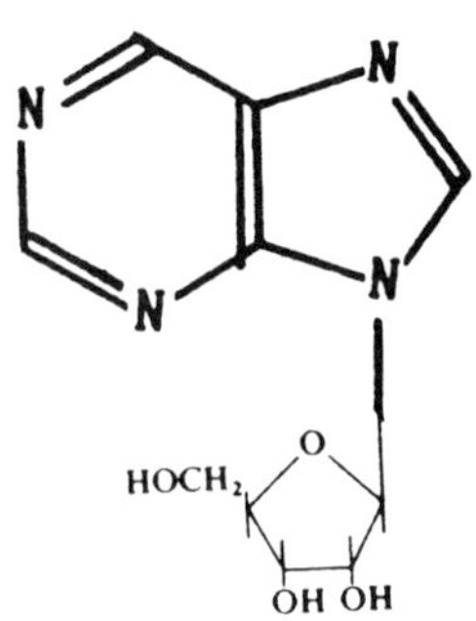

Pln

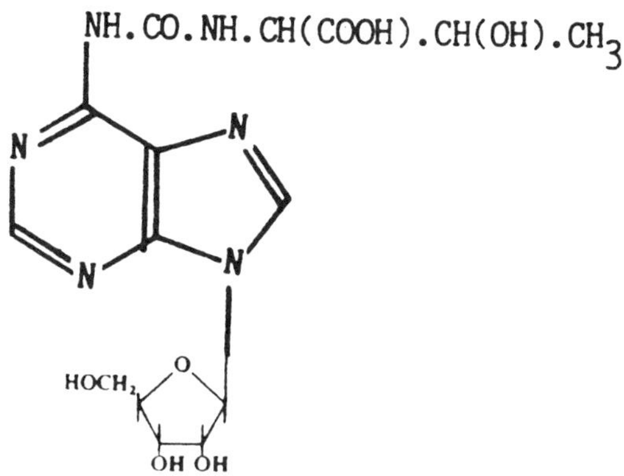
NH.CO.NH.CH(COOH).CH(OH).CH3
N
N
N
N
HOCH2
O
OH OH

P2n

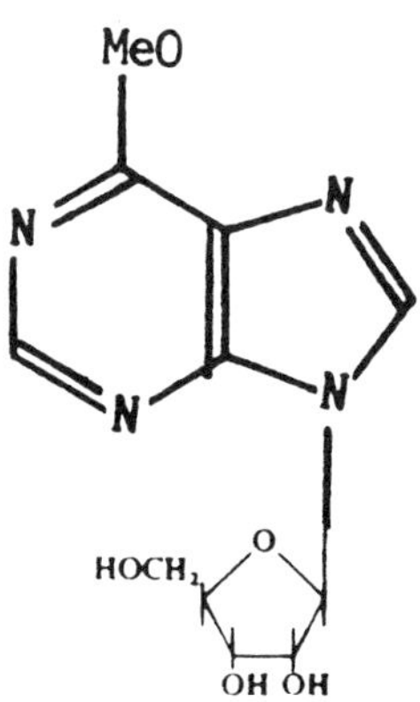
MeO
N
N
N
N
HOCH2
O
OH OH

P5n

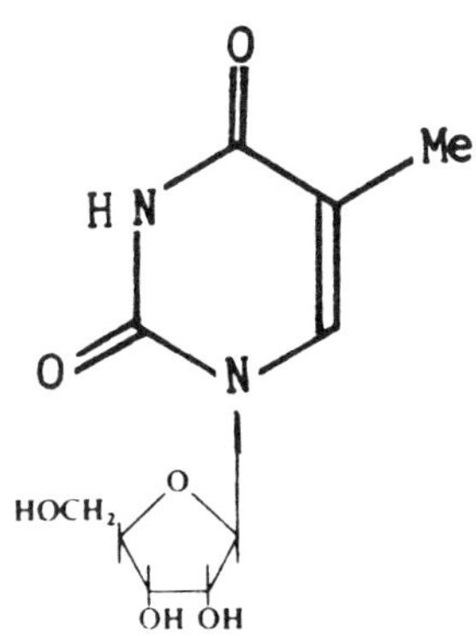
O
Me
H N
O
N
HOCH2
O
OH OH

T1n

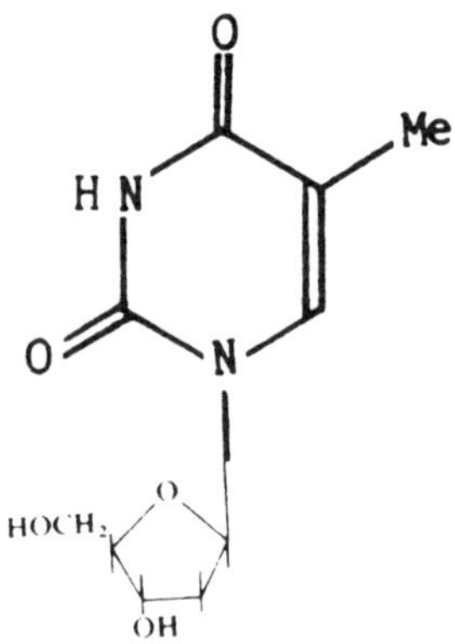

T3n

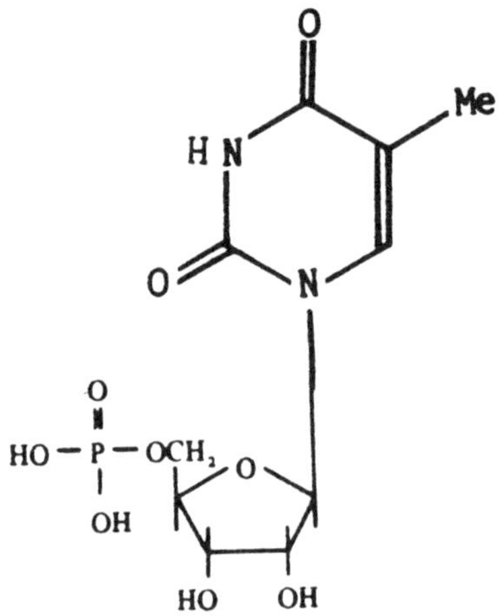

T1na

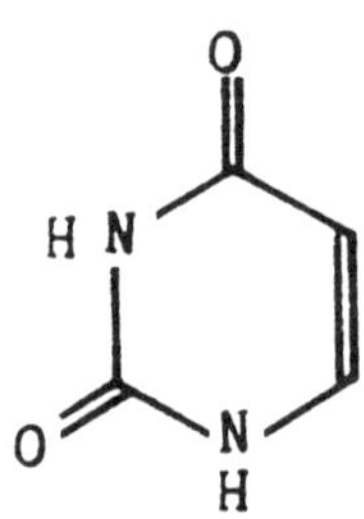

U1

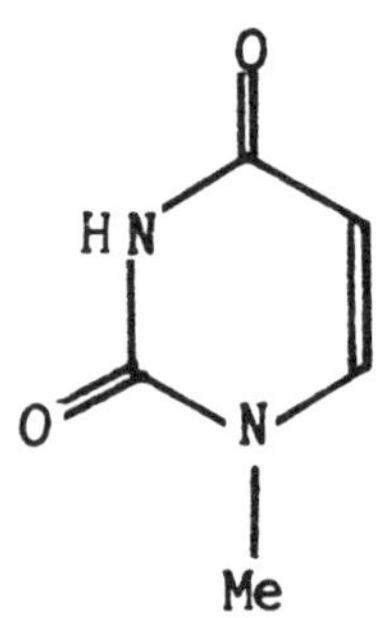

U2

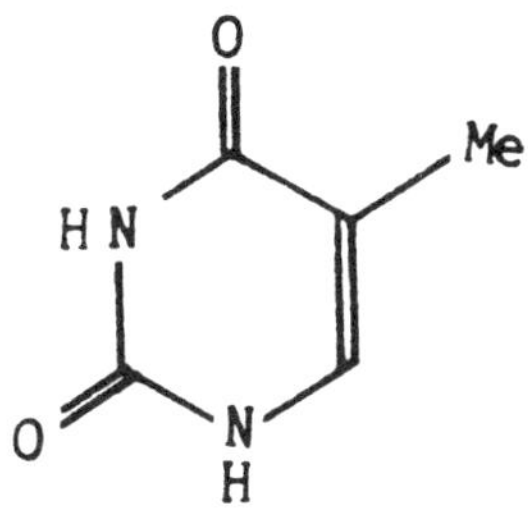

U3

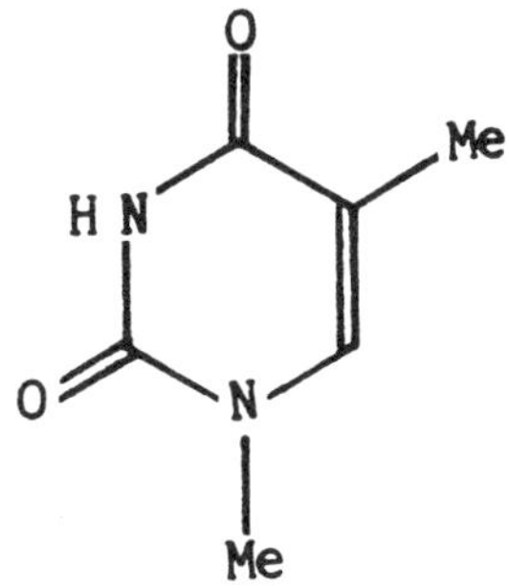

U4

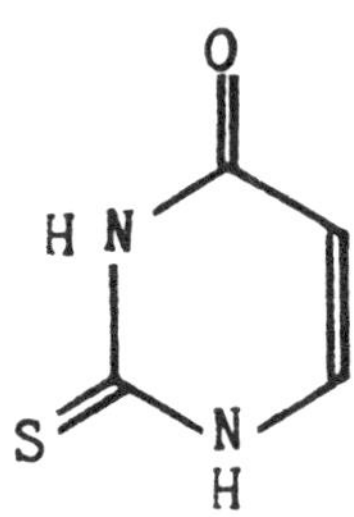

U5

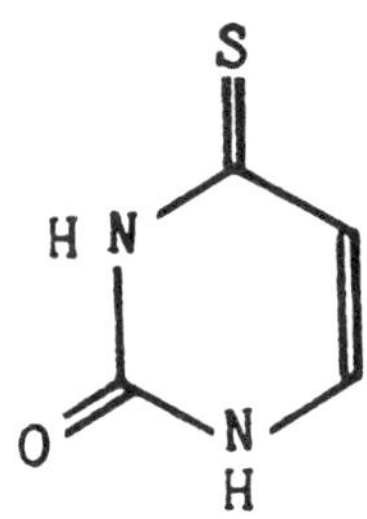

U7

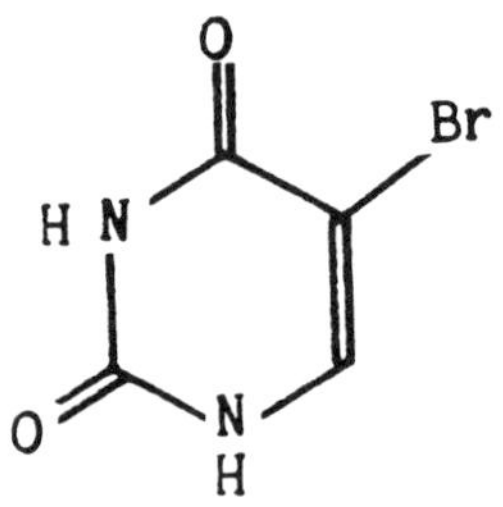

U8

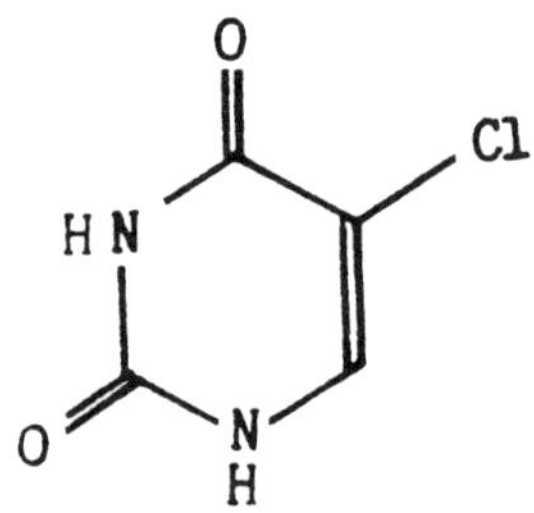

U9

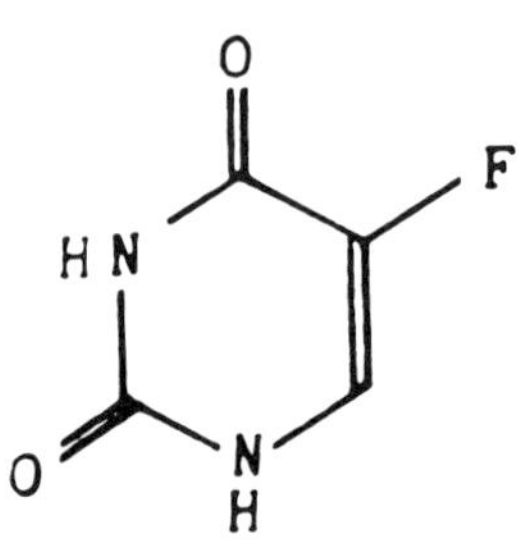

U10

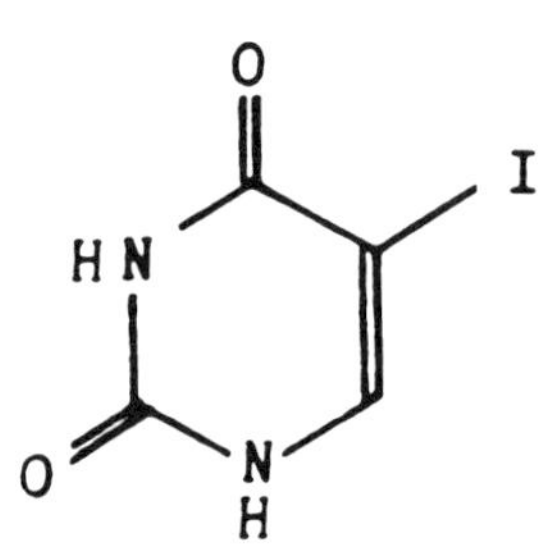

U11

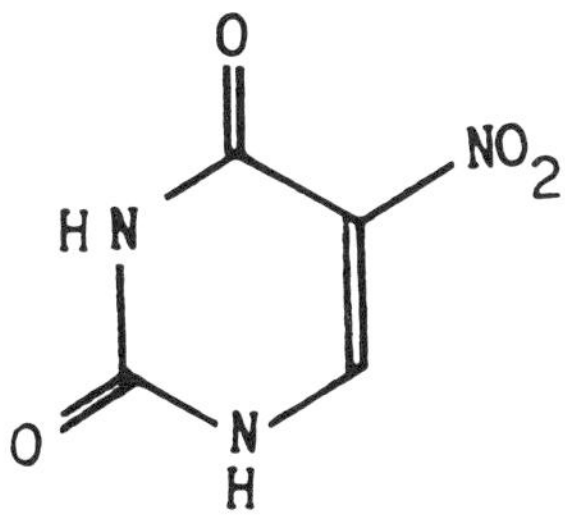

U12

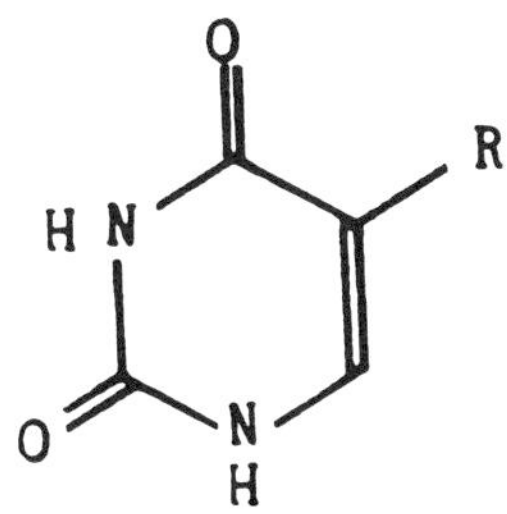

U13

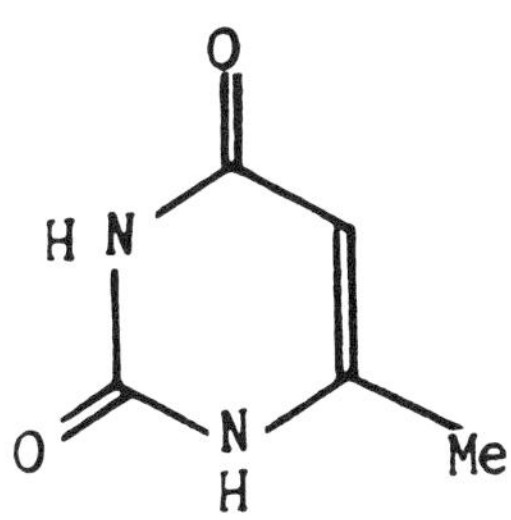

U14

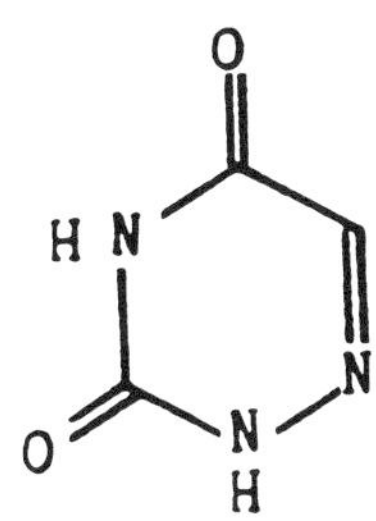

U15

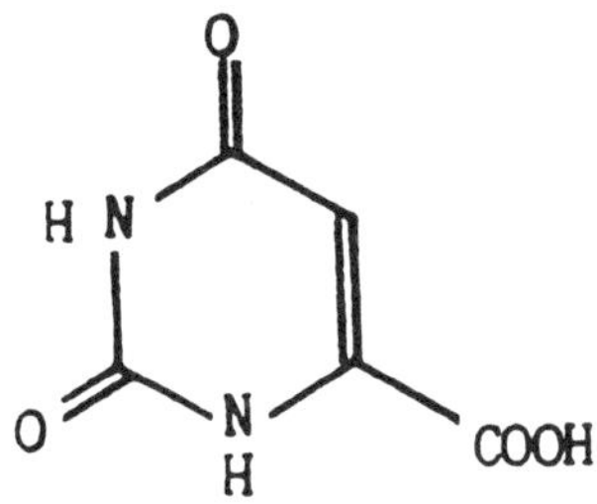

U16

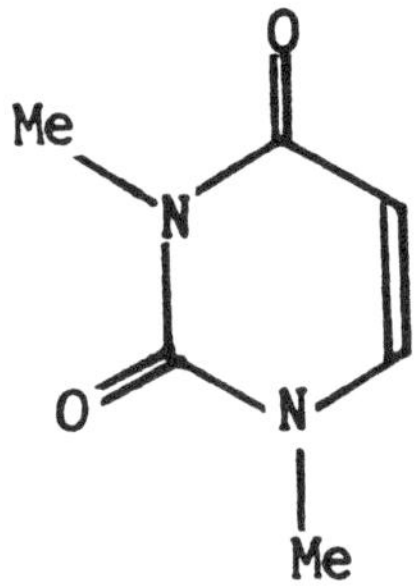

U17

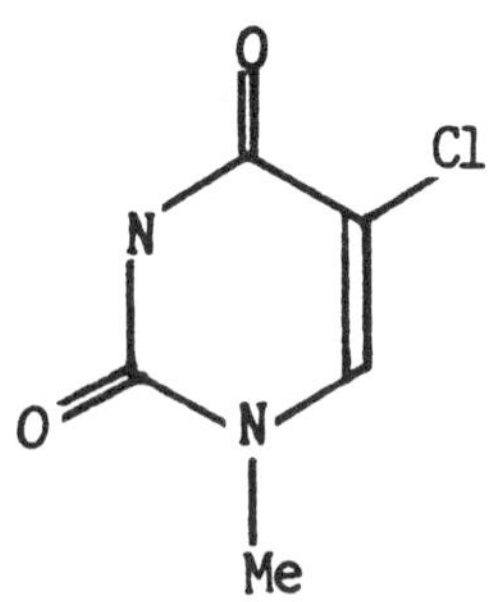

U18

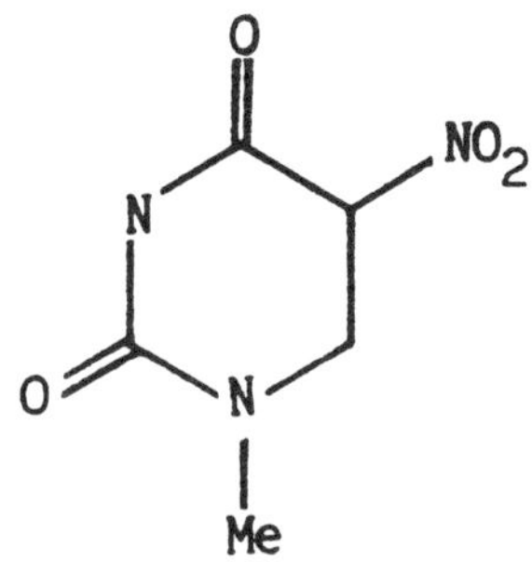

U19

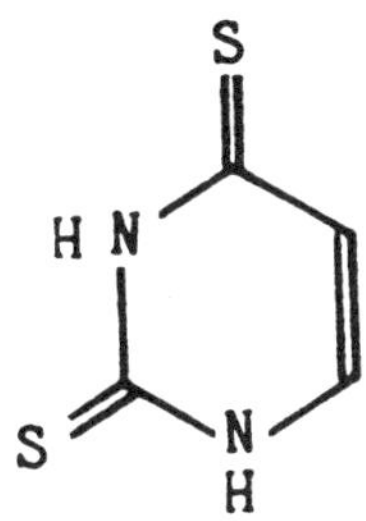

U20

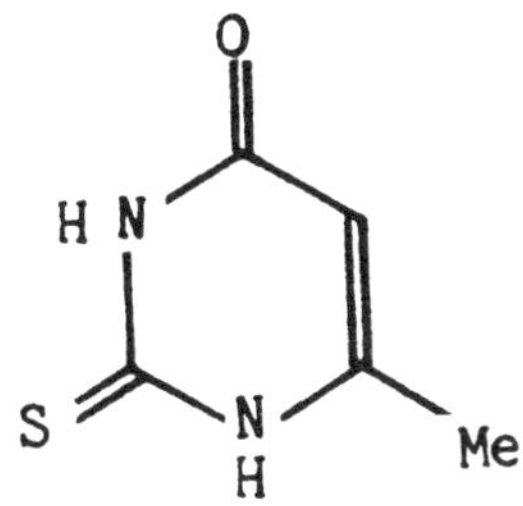

U21

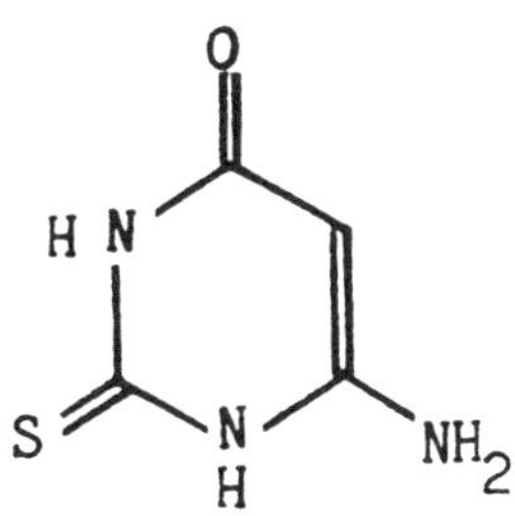

U22

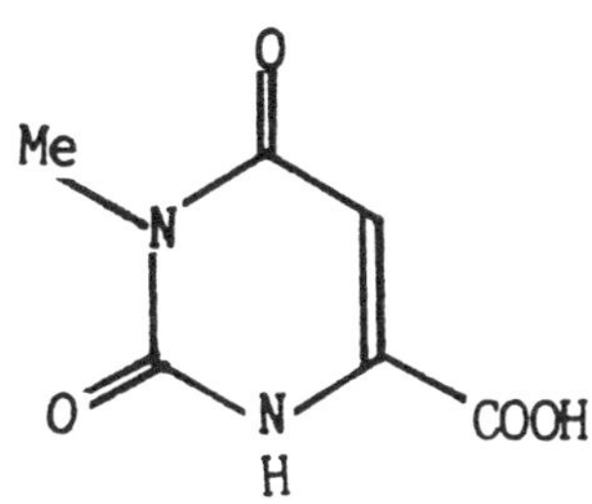

U23

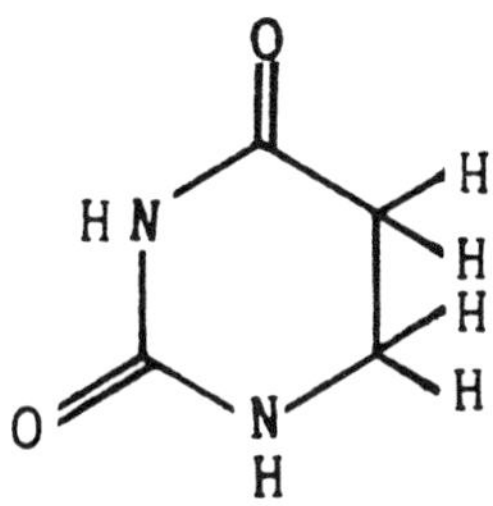

U24

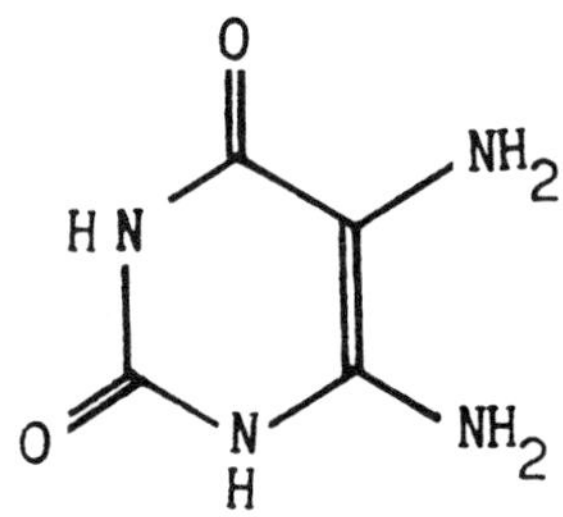

U25

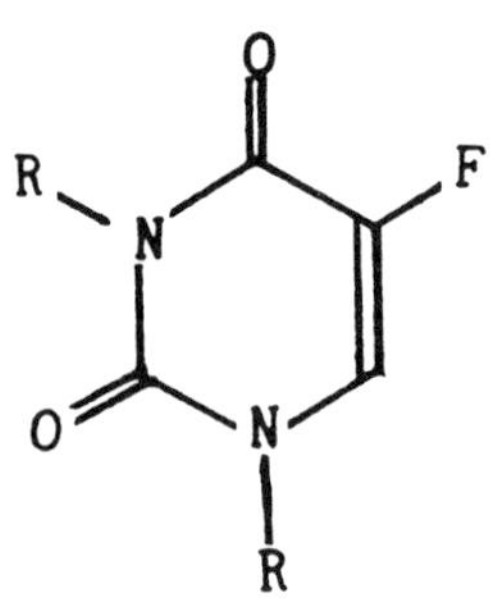

U33

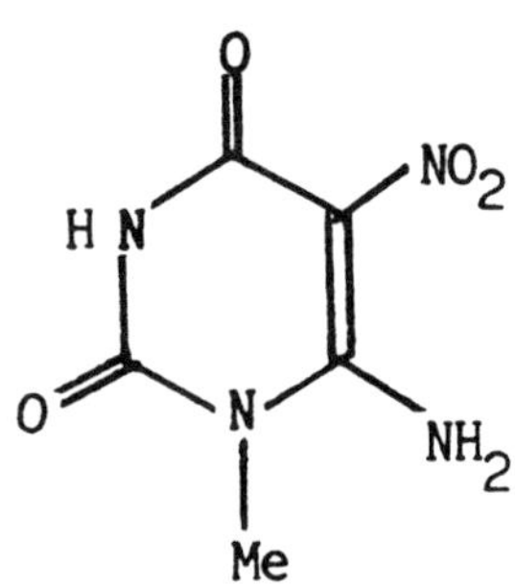

U34

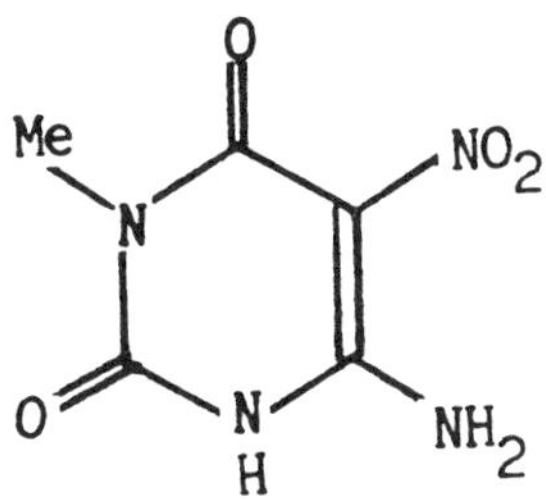

U35

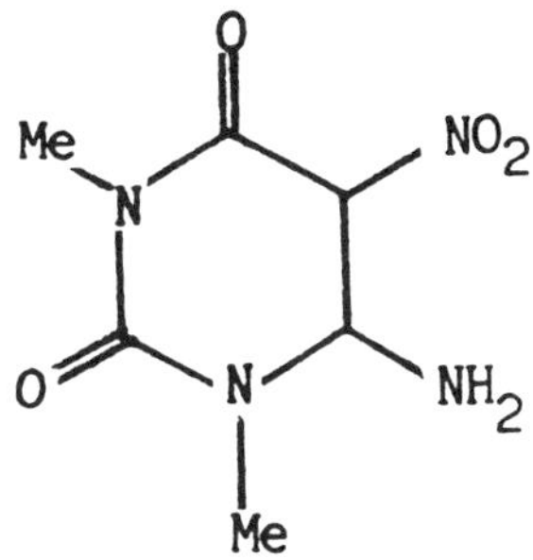

U36

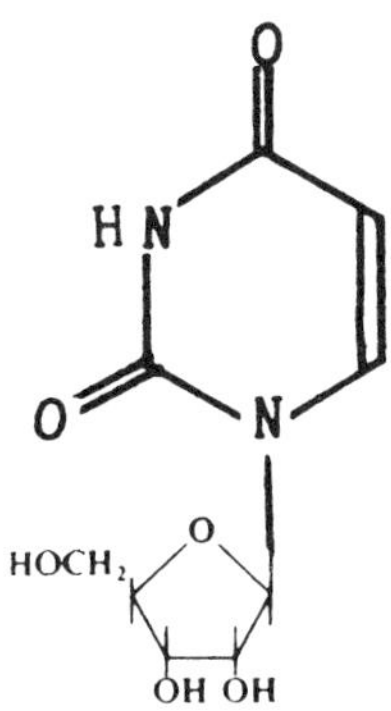

U1n

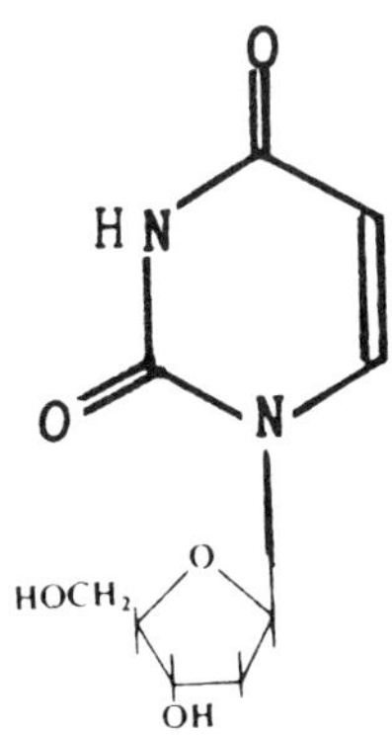

U2n

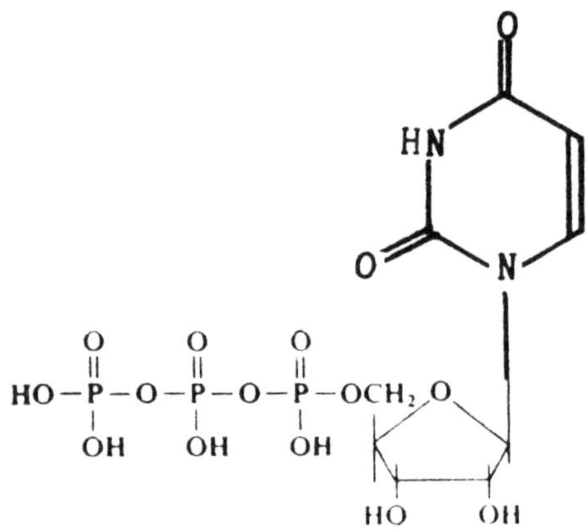

Ulnb

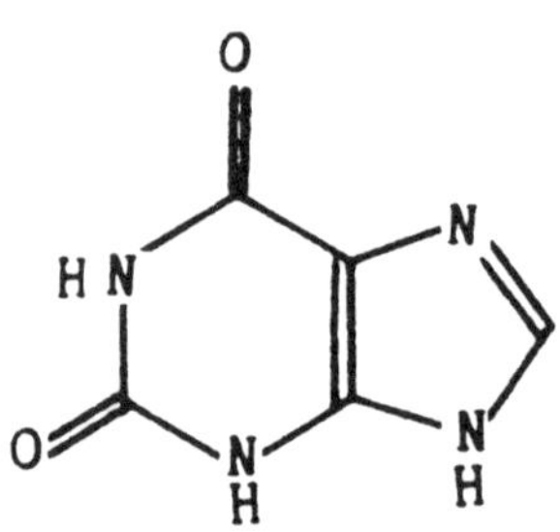

X1

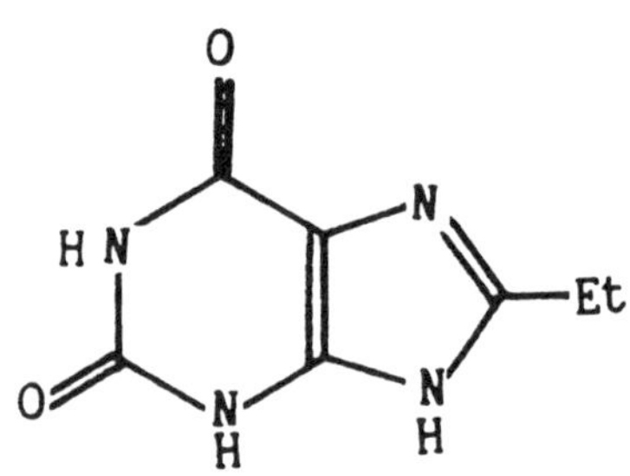

X2

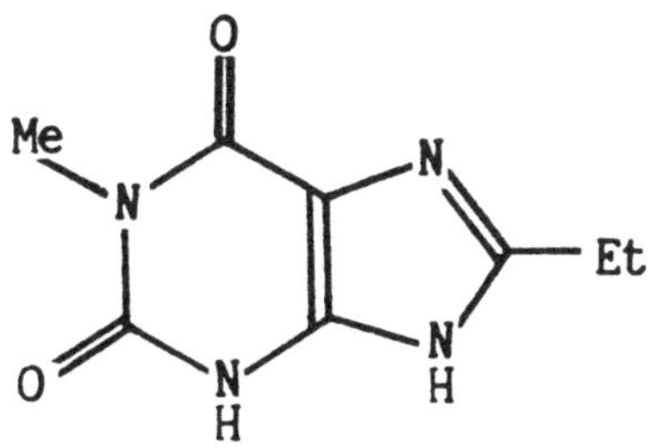

X3

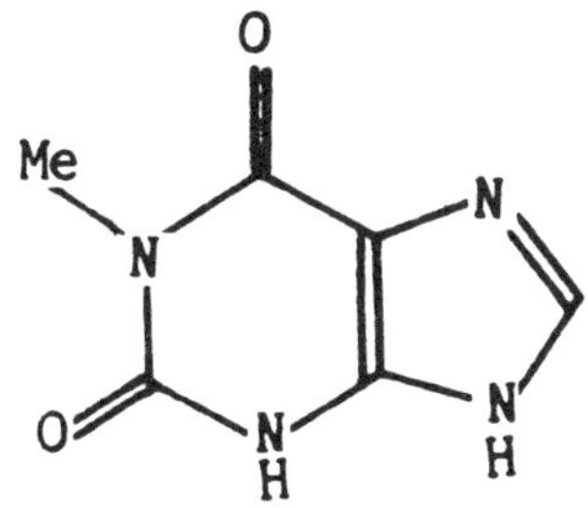

X5

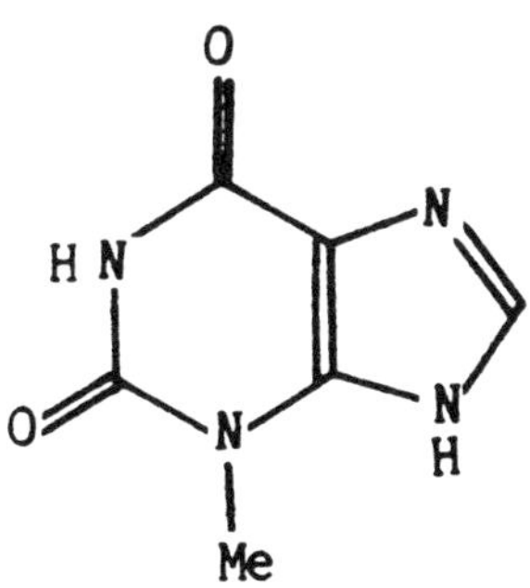

X6

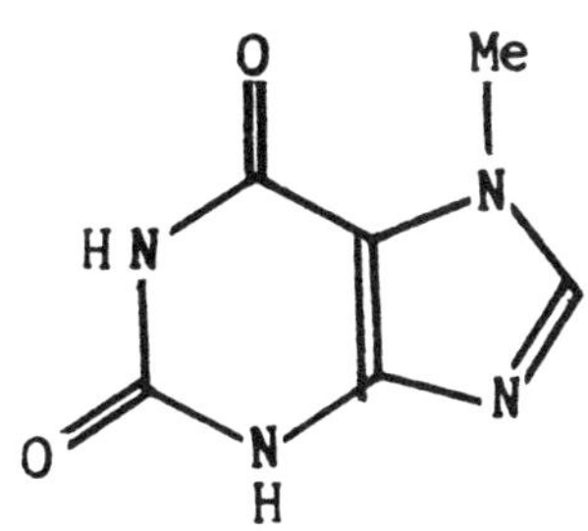

X7

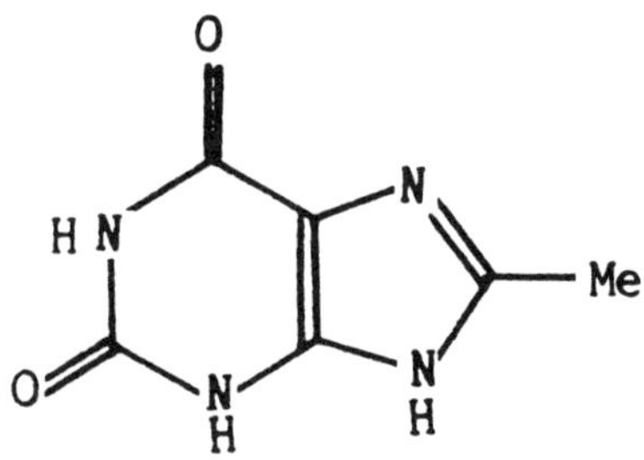

X8

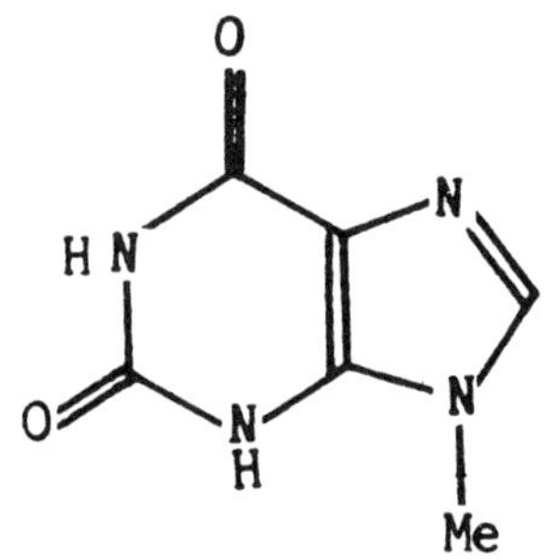

X9

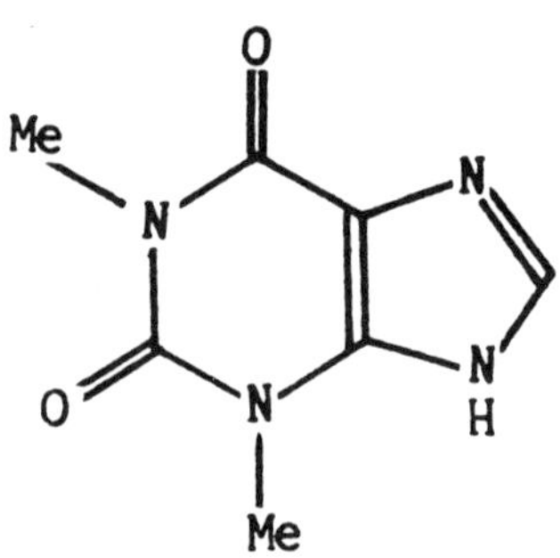

X10

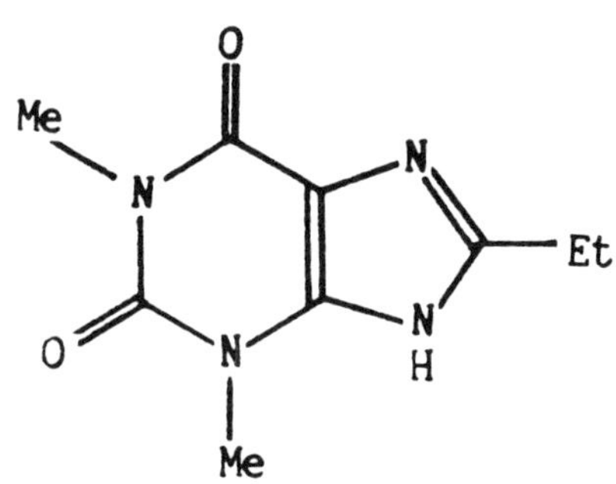

X11

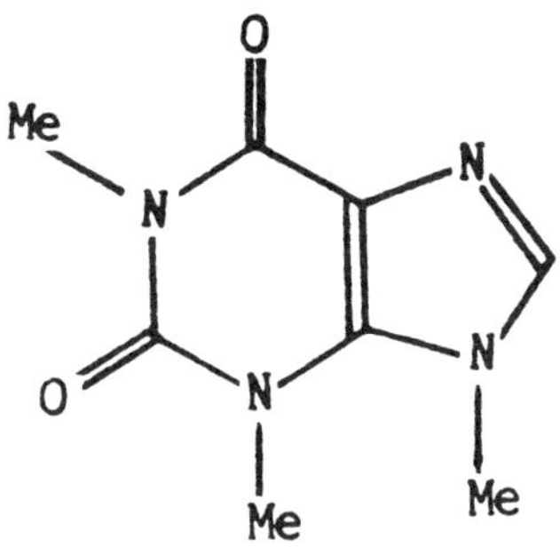

X12

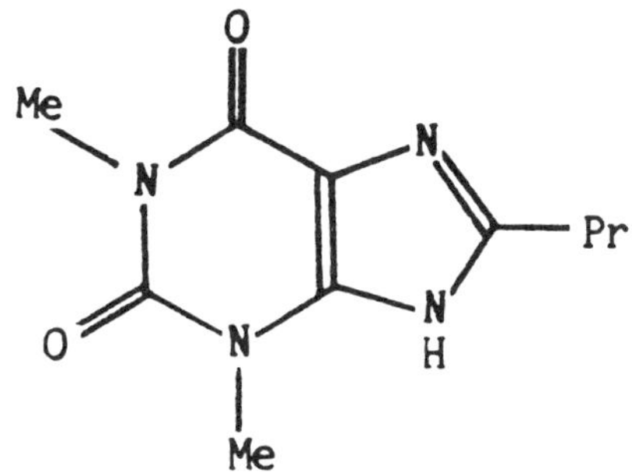

X15

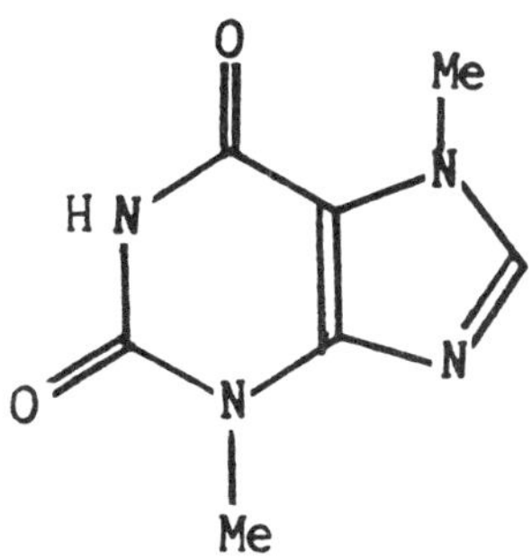

X20

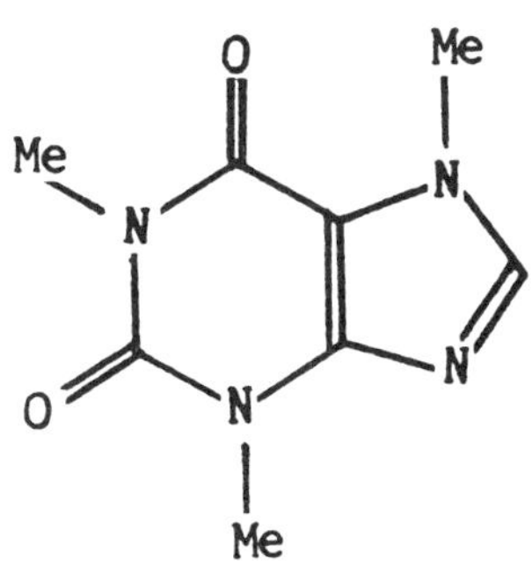

X30

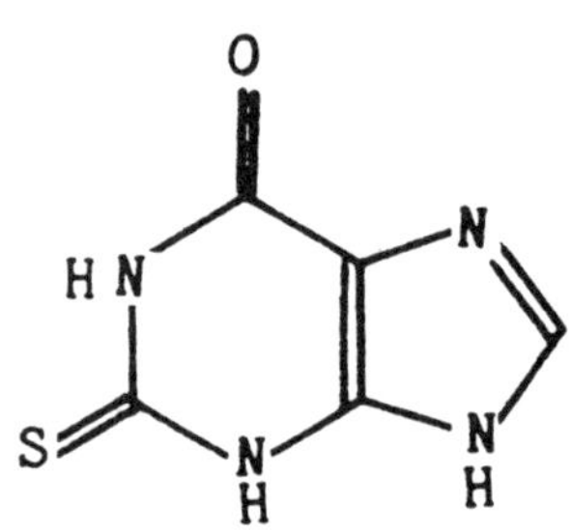

X37

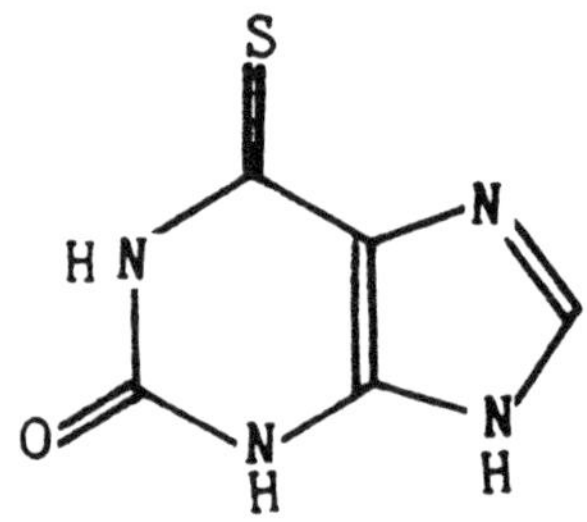

X38

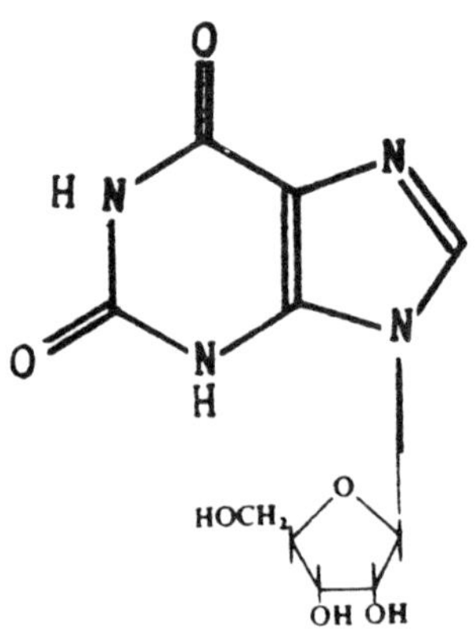

X1n

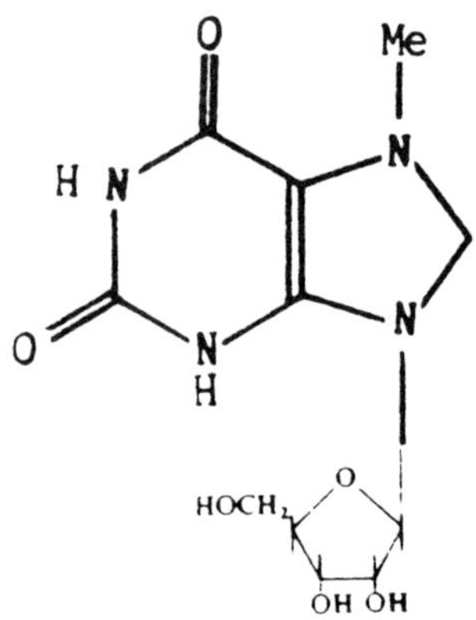

X2n

INDEX

A

B

C

D

E

F

G

H

I

M

N